AF333040

Springer Tracts in Modern Physics
Volume 216

Managing Editor: G. Höhler, Karlsruhe

Editors: A. Fujimori, Chiba
        C. Varma, California
        F. Steiner, Ulm
        J. Kühn, Karlsruhe
        J. Trümper, Garching
        P. Wölfle, Karlsruhe
        Th. Müller, Karlsruhe

Springer Tracts in Modern Physics

Springer Tracts in Modern Physics provides comprehensive and critical reviews of topics of current interest in physics. The following fields are emphasized: elementary particle physics, solid-state physics, complex systems, and fundamental astrophysics.

Suitable reviews of other fields can also be accepted. The editors encourage prospective authors to correspond with them in advance of submitting an article. For reviews of topics belonging to the above mentioned fields, they should address the responsible editor, otherwise the managing editor.

See also springer.com

Kirill Melnikov   Arkady Vainshtein

Theory of the Muon Anomalous Magnetic Moment

With 33 Figures

 Springer

Kirill Melnikov
University of Hawaii at Manoa
Department of Physics and Astronomy
Honolulu, HI 96822
USA
E-mail: kirill@phys.hawaii.edu

Arkady Vainshtein
William I. Fine Theoretical Physics Institute
School of Physics and Astronomy
University of Minnesota
116 Church Street SE
Minneapolis, MN 55455
USA
E-mail: vainshte@umn.edu

Library of Congress Control Number: 2006922615

Physics and Astronomy Classification Scheme (PACS):
11.00, 12.00, 13.40.Em, 14.60.Ef

ISSN print edition: 0081-3869
ISSN electronic edition: 1615-0430
ISBN-10   3-540-32806-8 Springer Berlin Heidelberg New York
ISBN-13   978-3-540-32806-3 Springer Berlin Heidelberg New York

Springer is a part of Springer Science+Business Media
springer.com
© Springer-Verlag Berlin Heidelberg 2006
Printed in The Netherlands

Typesetting: by the author using a Springer LATEX macro package
Cover concept: eStudio Calamar Steinen
Cover production: *design &production* GmbH, Heidelberg

Printed on acid-free paper   SPIN: 11013570   56/techbooks   5 4 3 2 1 0

Preface

Particle physics discoveries at the end of the 20th century confirmed an amazingly simple, yet powerful theory that describes fundamental interactions of elementary particles. This theory is usually referred to as the Standard Model of particle physics. It postulates the existence of three generations of quarks and leptons, four electroweak and eight strongly interacting gauge bosons and one scalar particle, the Higgs boson. Apart from the Higgs boson, the existence of all other particles has been unambiguously established. The Standard Model withstood numerous experimental tests during the last fifteen years, and has been tested at the level of precision sensitive to quantum loop corrections.

Given the fact that the Standard Model is so successful in describing Nature, is there any reason to believe that this theory is incomplete? There are different answers to this question, depending upon how seriously one takes certain theoretical prejudices. Most severe phenomenological problems appear when the Standard Model of particle physics is confronted with cosmological observations. For example, most of the matter in our Universe is in the form of a cold nonbaryonic dark matter that is not possible to describe within the Standard Model. The Standard Model has difficulties explaining observed baryon asymmetry in the Universe because the Cabbibo–Kobayashi–Maskawa mechanism of CP violation is too weak to produce a sufficient matter-antimatter asymmetry during the cosmological phase transition. Although neutrino oscillations can be easily incorporated into the Standard Model, such a solution is not considered theoretically appealing.

In addition to these problems rooted in *experimental facts*, there are shortcomings of the Standard Model of a theoretical nature. For example, it does not explain the fermion mass spectrum; there is a hierarchy and naturalness problem; gravity is not included in the Standard Model, and so on. Many theoretical ideas to address these issues have been proposed in the past; among them, the most prominent and well-developed is the "supersymmetry". Eventually, the experiment will decide which of these ideas are correct; luckily, a robust experimental program in particle physics, planned for the next decade, allows us to address these issues.

The Standard Model is studied experimentally at low and high energies. A typical experiment consists of a precise measurement of an observable and a comparison of the result of the measurement to a Standard Model prediction. At particle accelerators, one may hope to produce new particles directly if the energy of an accelerator is high enough. Whenever possible, this is a wonderful way to search for new physics and, once it is discovered, to explore it. On the other hand, searches for deviations caused by new physics in low-energy observables are complimentary to collider experiments. Such new physics effects are usually small; the essential part of any low-energy test is therefore a solid Standard Model prediction for a given observable.

The muon anomalous magnetic moment plays a special role in low-energy tests. It was first measured in the early sixties and, since then, several measurements of this quantity have been performed with increasing precision. The muon magnetic anomaly is interesting because (i) it can be precisely measured; (ii) it can be precisely computed in the Standard Model; and (iii) given the current level of precision on the muon anomalous magnetic moment, new physics with a mass scale of several hundred GeV can be detected.

It should be stressed, however, that both, measuring and computing the muon magnetic anomaly, becomes more and more difficult as the required precision increases. This happens because various subtle effects that could have been neglected earlier, start to play a role. On the theory side, this problem becomes especially acute when the muon magnetic anomaly calculation becomes sensitive to low-energy hadron physics, where our knowledge is limited.

It would not be much of an exaggeration to say that the physics of the muon anomalous magnetic moment is the contemporary particle physics in a nutshell. To arrive at the prediction sufficiently accurate for the current experimental effort, we require four-loop calculations in Quantum Electrodynamics, computation of two-loop electroweak effects in the Standard Model, analysis of the isospin symmetry breaking effects in hadronic decays of the τ lepton, precise measurement of the $e^+e^- \rightarrow$ hadrons annihilation cross-section, and an accurate estimate of the hadronic light-by-light scattering. Theoretical tools that are used for the analysis include advanced techniques for perturbative calculations, the operator product expansion in QCD, chiral perturbation theory, large-N_c QCD, renormalization group analysis and, since very recently, even lattice gauge theories. A discrepancy between the experimental result and the Standard Model prediction can be explained by invoking new physics effects that may include supersymmetry, extra dimensions, anomalous gauge boson interactions, modifications of the Higgs sector of the Standard Model, etc. Mastering the physics of the muon magnetic anomaly requires dealing with all these issues, which makes the anomalous magnetic moment quite an interesting observable from a theoretical viewpoint.

The purpose of this book is to provide a review of the physics of the muon anomalous magnetic moment. The time for such a review is perhaps appropriate because the most recent experiment at Brookhaven National Laboratory,

E821, has now been completed. Its legacy, the $2 - 3\ \sigma$ discrepancy between the experimental and the theoretical (Standard Model) results for the muon magnetic anomaly is inconclusive but intriguing. In spite of the fact that the final results of E821 do not tell us if the physics beyond the Standard Model is required to understand the muon magnetic anomaly, in recent years this experiment stimulated a great deal of theoretical studies of the muon anomalous magnetic moment. Our goal here is to summarize them. We try to do that in a pedagogical, yet thorough fashion. The theory of the muon magnetic anomaly has reached a new stage in recent years; we hope that by reviewing it, we might contribute to stimulating further progress.

A few words about the layout of the book. Besides the Introduction and the Summary, it contains eight chapters that discuss various contributions to the muon magnetic anomaly. Each chapter deals with a specific contribution to the muon anomalous magnetic moment, and hence, can be read independently.

We begin by reviewing the QED contribution to the muon anomalous magnetic moment in Chap. 2. Then, in Chap. 3, we deal with the hadronic vacuum polarization component of the muon magnetic anomaly. In that chapter, we introduce theoretical methods such as large-N_c approximation in QCD as well as the operator product expansion, which are used extensively throughout the book. An important (and controversial) topic in recent studies of hadronic vacuum polarization contribution to the muon magnetic anomaly is the use of data on hadronic decays of τ lepton. We provide a detailed discussion of the relation between the data on $e^+e^- \to$ hadrons and $\tau \to \nu_\tau +$ hadrons in Chap. 3.

In Chap. 4 we discuss electroweak corrections to the muon magnetic anomaly. We show that some of those corrections are sensitive to nonperturbative hadronic effects. A more careful discussion of hadronic effects in weak corrections to the muon anomalous magnetic moment is given in Chap. 5. Calculation of the hadronic light-by-light scattering contribution to the muon magnetic anomaly is described in Chap. 6.

In Chap. 7 the Standard Model prediction for the muon anomalous magnetic moment is given and its relation to other precision observables is discussed. We show that the Standard Model value for the muon magnetic anomaly differs from the experimental result by 2–3 standard deviations. A possible explanation of this discrepancy beyond the Standard Model physics is discussed in Chap. 8.

The notations used throughout the text are summarized in Appendix A.1.

Acknowledgments

We are grateful to A. Czarnecki, M. Davier, A. Höcker, W. Marciano, M. Shifman, and M. Voloshin for numerous discussions of the physics of the muon anomalous magnetic moment. We would like to thank S. Eidelman and M. Voloshin for comments on the manuscript. Partial support by U.S. Department of Energy under contracts DE-FG03-94ER-40833 and DEFG02-94ER408 and the Alfred P. Sloan Foundation is gratefully acknowledged.

Kirill Melnikov
Arkady Vainshtein

March 2006

Contents

1 Introduction

1.1 Preliminary Remarks

The motion of the classical particle with the angular momentum $\boldsymbol{L} = \boldsymbol{r} \times \boldsymbol{p}$ generates the magnetic moment $\boldsymbol{\mu}$,

$$\boldsymbol{\mu} = \frac{e}{2mc}\,\boldsymbol{L}\,,\tag{1.1}$$

where e is the charge of the particle and m is its mass. In quantum mechanics, the angular momentum $\boldsymbol{L}$ becomes an operator, $\boldsymbol{L} = \hbar\boldsymbol{l} = -i\hbar\,\boldsymbol{r} \times \boldsymbol{\nabla}$, whose eigenvalues are quantized in units of the Planck constant $\hbar$. The magnetic moment associated with the orbital motion is quantized accordingly.

Besides the orbital motion, an elementary particle may have internal rotation characterized by the spin $\boldsymbol{S} = \hbar\boldsymbol{s}$. The magnetic moment associated with this rotation can be presented in the form similar to (1.1),

$$\boldsymbol{\mu} = g\,\frac{e\hbar}{2mc}\,\boldsymbol{s}\,,\tag{1.2}$$

with an additional factor g which is called the gyromagnetic factor. While for the orbital motion this factor is equal to 1, the Dirac equation for a charged elementary fermion with spin $s = 1/2$ implies $g = 2$. The anomalous magnetic moment refers to a deviation of the gyromagnetic factor from the $g - 2$ value and is parametrized by $a = (g-2)/2$. It appears due to radiative corrections. The leading contribution to a, calculated by Schwinger in 1949,

$$a = \frac{\alpha}{2\pi}\,,\tag{1.3}$$

is the same for the electron and the muon. Higher order effects are not universal and are affected by weak and strong interactions in addition to electromagnetism.

Recent results [1] from the experiment E821 at Brookhaven National Laboratory (BNL) might indicate a disagreement between the experimental value for the muon magnetic anomaly and the theoretical expectation based on the Standard Model. There are two experimental results from E821 with the precision better than one part per million. They are

K. Melnikov and A. Vainshtein: *Theory of the Muon Anomalous Magnetic Moment*
STMP **216**, 1–6 (2006)
© Springer-Verlag Berlin Heidelberg 2006

$$a_{\mu^+} = 116\,592\,039(84) \times 10^{-11}\,,$$
$$a_{\mu^-} = 116\,592\,143(83) \times 10^{-11}\,.$$

$$(1.4)$$

The errors are dominated by statistical uncertainties and hence can be further reduced, in principle.

Theoretical predictions for a_μ differ from these experimental results by an amount that is larger than the experimental uncertainty. Depending on the details on how corrections to Schwinger's formula (1.3) are implemented, the theoretical prediction for a_μ may disagree with the experimental results (1.4) by up to three standard deviations. Leaving the discrepancy aside for the time being, we stress that the precision of the theoretical prediction is very high. The current uncertainty in a_μ^{theory} is of the order of 100×10^{-11}; this implies that the Standard Model (SM) prediction for a_μ is accurate at the level of one part per million, an extraordinary result by itself. Achieving such a precision required advances in perturbative calculations in QED and electroweak physics and improvements in understanding the non-perturbative hadronic effects that influence the muon anomalous magnetic moment.

To a large extent, both QED and electroweak corrections to a_μ are currently understood at the level required by the experimental precision. The major contribution to the theoretical uncertainty comes from hadronic component of the muon magnetic anomaly. There are two distinct parts of the hadronic contribution – the hadronic vacuum polarization and the hadronic light-by-light scattering. Currently, they enjoy different level of understanding and, as a consequence, require somewhat different expertise for further improvements.

As we discuss in Chap. 3, hadronic vacuum polarization contribution to the muon anomalous magnetic moment is large, $\sim 7000 \times 10^{-11}$. Comparison with the current uncertainty of about 100×10^{-11} shows that calculations of the hadronic vacuum polarization should be controlled at the level of one percent.

The hadronic vacuum polarization contribution to the muon magnetic anomaly has traditionally been computed by using data on e^+e^- annihilation into hadrons, accumulated over many years at different accelerators worldwide. More recently, it was suggested to incorporate data on hadronic decays of the τ lepton, $\tau \to \nu_\tau + X_{\text{hadr}}$, collected and analyzed by ALEPH and OPAL collaborations at LEP and by CLEO collaboration at CESR. The gist of this idea is that the decay rate of the τ lepton and the e^+e^- annihilation cross-section for a given invariant mass of the produced hadronic state are related by the isospin symmetry. While the isospin symmetry is violated by both the mass and the electric charge differences of up and down quarks, this effect is small and confronting the analyses based on the e^+e^- and the τ data helps in elucidating potential deficiencies. However, the benefits of incorporating the τ data are less clear if reaching high precision is required. Then, the isospin violation correction has to be applied to the τ data; unfortunately, as we discuss in Chap. 3, computation of such corrections is a non-trivial problem by itself.

The situation with the hadronic light-by-light scattering is different. Although this contribution is numerically small, we have to rely on theoretical considerations for its evaluation since there is no data on low-energy photon-photon scattering. As we discuss in Sect. 6, model-independent approaches to the hadronic light-by-light scattering based on, e.g., chiral perturbation theory, are insufficient; as a consequence, some degree of model dependence of theoretical calculations can not be avoided. Typically, such a modeling involves an approximate description of strong interactions. Unfortunately, it is difficult to improve on the current situation given a limited understanding of low-energy hadron physics. As a consequence, the hadronic light-by-light scattering can become a principal obstacle for further improvements in precision of the theoretical description of the muon magnetic anomaly.

As we mentioned earlier, currently there is a $2 - 3\,\sigma$ discrepancy when experimental results are confronted with theoretical expectations. By itself, this is not very exciting since in the history of particle physics many 2σ discrepancies appeared only to be resolved later by mundane explanations. It is conceivable that the discrepancy in the muon magnetic anomaly will follow the same fate. However, if the discrepancy persists and no explanation within the Standard Model is found, our last resort – the physics beyond the Standard Model – is invoked. Quite generally, new physics associated with the energy scale Λ is expected to change the anomalous magnetic moment of the muon by [1]

$$
a_\mu^{\rm BSM} \sim \left(\frac{\alpha}{\pi} \right) \frac{m^2}{\Lambda^2} \approx 230 \left(\frac{100\ {\rm GeV}}{\Lambda} \right)^2 \times 10^{-11} \, ,
\tag{1.5}
$$

Taking $a_\mu^{\rm BSM}$ to be of the same order as the current experimental uncertainty, $\sim 100 \times 10^{-11}$, we find that the muon anomalous magnetic moment probes $\Lambda \sim 100$ GeV. This is exactly the range of energies where particle physicists expect new phenomena to occur. The order of magnitude estimate (1.5) is corroborated in Chap. 8 where various scenarios of beyond the Standard Model physics are discussed in connection with the muon magnetic anomaly.

Since the muon anomalous magnetic moment has been studied for, approximately, fifty years, there is substantial body of literature devoted to theoretical and experimental aspects of the physics of the muon magnetic anomaly. Reviews on the subject, both old and more recent, can be found in [2, 3, 4, 5, 6, 7, 8, 9].

---

[1] The notations used throughout the text are summarized in Appendix A.1.

1.2 Muon Spin Precession in the Storage Ring

Let us recall how the muon anomalous magnetic moment is measured in contemporary experiments. Our discussion here is sketchy; for more details we refer to two reviews [3, 4] were all the experiments on the muon magnetic anomaly are discussed in great detail.

If the muon is placed in a magnetic field $\boldsymbol{B}$, the muon spin precesses around the direction of this field. The spin precession frequency is given by

$$\boldsymbol{\omega}_{\mathrm{spin}} = -g\,\frac{e}{2mc}\,\boldsymbol{B}\,. \tag{1.6}$$

Provided that all entries but g in (1.6) are known precisely, measuring the spin precession frequency gives us the muon g-factor. Since $g = 2 + 2a_\mu$ and $a_\mu \sim 10^{-3}$, measuring a_μ to a relative precision of 10^{-6} requires measuring g to *one part per billion*, which is an utopian goal. Note that in this discussion we completely ignored the fact that a muon is an unstable particle with a short lifetime; accounting for that, introduces additional complications.

For a high precision measurement of the muon magnetic anomaly, we must measure a_μ *directly*; this would give us an improvement in sensitivity by three orders of magnitude right from the start. A special feature of the Dirac equation comes to rescue: a projection of the spin on the velocity of the particle, $\boldsymbol{s}(\boldsymbol{p} - (e/c)\boldsymbol{A})$, is conserved. This leads to the observation [10] that the cyclotron frequency of the muon in a magnetic field coincides with the muon spin precession frequency if $a_\mu = 0$. Hence, if the muon moves along a circular orbit in a magnetic field, the angle between its velocity and spin oscillates with the frequency

$$\boldsymbol{\omega}_a = -a_\mu\,\frac{e}{mc}\,\boldsymbol{B}\,, \tag{1.7}$$

which enables the direct measurement of a_μ.

To have a realistic experimental method, we have *i*) to know the initial direction of the muon spin; *ii*) to ensure that the experiment allows for sufficiently many spin oscillations during the muon lifetime; and *iii*) to find a way to analyze the direction of the muon spin at the end of the measurement. Solutions to those problems were suggested by Farley in 1961 [11]; this heralded the beginning of the contemporary era in experimental studies of the muon magnetic anomaly.

The idea of the method is easy to explain. Colliding a proton beam with a hadronic target creates energetic pions that are captured in a storage ring. The pions decay into muons; the muons remain in the storage ring. The muons are relativistic and hence live long; in BNL experiment, a typical muon life time is ~ 60 μs. Since muons from pion decays are polarized, at $t = 0$ the directions of the muon spin and the muon momentum coincide. The clock stops with the decay of the muon; at that moment we want to know the direction of the muon spin. For this purpose, we use the property of the muon decay; the $V - A$ structure of the charged weak current ensures that electrons from the muon decay $\mu \to e\bar{\nu}_e\nu_\mu$ prefer to follow the direction opposite to the direction of the muon spin. In the laboratory frame, the direction of the muon spin oscillates with the frequency ω_a; therefore, the number of electrons arriving into the solid angle $\mathrm{d}\Omega$ as a function of time is

$$\frac{\mathrm{d}N_e(t)}{\mathrm{d}\Omega} \sim e^{-\Gamma t}\left(1 + A\cos(\omega_a t + \phi)\right). \tag{1.8}$$

Hence, the number of electrons that are registered by a detector positioned at a given angle exhibits modulated oscillations with the frequency ω_a. By measuring these oscillations, we determine the frequency ω_a and, using (1.7), the muon anomalous magnetic moment.

The above discussion provides the gist of the idea behind the contemporary measurements of the muon anomalous magnetic moment. However, in an experiment that aims at the precision of one part per million, many subtle issues have to be carefully addressed. Some of those issues are discussed below; further details can be found in [3, 4].

The muons are injected into the storage ring with different momenta; to keep them in a plane transverse to the magnetic field, the quadrupole electric field $\boldsymbol{E}$ is applied. However, the electric field interacts differently with the muon spin and the muon orbital momentum and therefore changes the spin precession frequency,

$$\boldsymbol{\omega}_a = -a_\mu \frac{e}{mc}\boldsymbol{B} - \frac{e}{mc^2}\left(\frac{1}{\gamma^2 - 1} - a_\mu\right)\boldsymbol{v} \times \boldsymbol{E}. \tag{1.9}$$

In this formula, $\boldsymbol{v}$ is the muon velocity and $\gamma = E_\mu/m$, where E_μ is the energy of the muon. The last term in (1.9) would have had disastrous consequences for measuring a_μ since it would require a very precise knowledge of the electric field $\boldsymbol{E}$. Fortunately, one can cancel this term altogether, by choosing the appropriate energy for the muons. Indeed, if the muon energy is such that $(\gamma^2 - 1)^{-1} = a_\mu$, the term proportional to the electric field in (1.9) vanishes. Numerically, the "magic" energy works out to be

$$E_\mu = m\sqrt{\frac{1}{a_\mu} + 1} \approx 3~\text{GeV}. \tag{1.10}$$

Once the spin precession frequency ω_a is measured, the value of the anomalous magnetic moment has to be extracted. To do so, we need to know the magnetic field with the precision of one part per million. This is achieved by measuring the *proton* spin precession frequency with nuclear magnetic resonance technique and using the ratio of the muon to proton magnetic moments measured in experiments with muonium [12]. Let us stress that if the hyperfine structure of muonium has not been measured with the required precision, there is no way to extract the value of the muon anomalous magnetic moment from measured spin precession frequency. This is an interesting connection of the frontier experiment in particle physics with experiments in atomic spectroscopy, often considered rudimentary by the high-energy physics community.

Because CP symmetry is broken, elementary particles may have, in addition to the magnetic moment, the electric dipole moment. In the Standard Model CP symmetry is violated by the Cabbibo–Kobayashi–Maskawa mixing matrix in the quark sector. The discovery of neutrino oscillations makes it conceivable that CP violation exists in the lepton sector of the Standard Model as well. Supersymmetry provides an explicit realization of this possibility.

The presence of the electric dipole moment modifies equations of motion for the muon spin in two different ways [13]. First, the spin precession frequency increases and, second, the component of the spin precession frequency along the $v \times B$ axis appears. Because of that, the electric dipole moment leads to the appearance of the "out of plane" component of the muon spin precession that can be measured if detectors are placed above and below the storage plane. A new dedicated experiment to search for the muon electric dipole moment is now being planned at J-PARC [14].

References

1. H. N. Brown et al. [Muon g-2 Collaboration], Phys. Rev. D **62**, 091101 (2000); Phys. Rev. Lett. **86**, 2227 (2001).
   G. W. Bennett et al. [Muon g-2 Collaboration], Phys. Rev. Lett. **89**, 101804 (2002); Erratum-ibid. **89**, 129903 (2002).
   G. W. Bennett et al. [Muon g-2 Collaboration], Phys. Rev. Lett. **92**, 161802 (2004).
2. T. Kinoshita (ed), *Quantum Electrodynamics*, World Scientific, 1990.
3. F. J. M. Farley and E. Picasso, *The muon $g - 2$ experiments*, in [2].
4. F. J. M. Farley and Y. K. Semertzidis, Prog. in Part. and Nucl. Phys. **52**, 1 (2004).
5. B. E. Lautrup, A. Peterman and E. de Rafael, Phys. Rept. **3**, 193 (1972).
6. J. Calmet, S. Narison, M. Perrottet and E. de Rafael, Rev. Mod. Phys. **49**, 21 (1977).
7. T. Kinoshita and W. J. Marciano, *Theory of the muon anomalous magnetic moment*, in [2].
8. M. Knecht, Lect. Notes Phys. **629**, 37 (2004) [hep-ph/0307239].
9. M. Passera, J. Phys. G **31**, 75 (2005).
10. W. H. Louisell, R. W. Pidd and H. R. Crane, Phys. Rev. **91**, 475 (1953); A. A. Schupp, R. W. Pidd and H. R. Crane, Phys. Rev. **121**, 1 (1961).
11. F. J. M. Farley, *Proposed high precision $(g-2)$ experiment*, CERN Intern. Rep. NP/4733 (1962).
12. W. Liu et al., Phys. Rev. Lett. **82**, 711 (1999).
13. J. Feng, K. Matchev and Y. Shadmi, Phys. Lett. B **555**, 89 (2003).
14. J-PARC letter of Intent L22, *Search for a permanent muon electric dipole moment at the 10^{-24} e cm level*, Y. Kuno, J. Miller and Y. Semertzidis (spokespersons).

2 QED Effects in the Muon Magnetic Anomaly

2.1 General Considerations

In the context of Quantum Electrodynamics (QED), interactions of muons and photons are described by the Lagrangian

$$\mathcal{L} = -\frac{1}{4}F_{\mu\nu}F^{\mu\nu} + \bar{\psi}\left(i\hat{\partial} - m\right)\psi - eJ^\mu A_\mu \;, \tag{2.1}$$

where $\psi(x)$ is the muon field, $A^\mu = (\varphi, \boldsymbol{A})$ is the vector potential of the electromagnetic field, $F^{\mu\nu} = \partial^\mu A^\nu - \partial^\nu A^\mu$ is the field-strength tensor of the electromagnetic field, $J^\mu(x) = \bar{\psi}(x)\gamma^\mu\psi(x)$ is the electric current and $e = -|e|$ is the muon electric charge.

Consider a situation when the muon with four momentum p_1 scatters off the external electromagnetic potential A_μ, Fig. 2.1. The momentum of the muon in the final state is p_2. To first order in the external field, the interaction is described by the scattering amplitude

$$\mathcal{M} = -ie\langle \mu_{p_2}|J^\mu(0)|\mu_{p_1}\rangle A_\mu(q) \;, \tag{2.2}$$

where $q = p_2 - p_1$. Thanks to current conservation $\partial_\mu J^\mu(x) = 0$ and parity conservation in QED, the most general parametrization of the matrix element reads

$$\langle \mu_{p_2}|J^\mu(0)|\mu_{p_1}\rangle = \bar{u}_{p_2}\Gamma^\mu(p_2,p_1)u_{p_1} = \bar{u}_{p_2}\left[F_D(q^2)\gamma^\mu + F_P(q^2)\frac{i\sigma^{\mu\nu}q_\nu}{2m}\right]u_{p_1} \;, \tag{2.3}$$

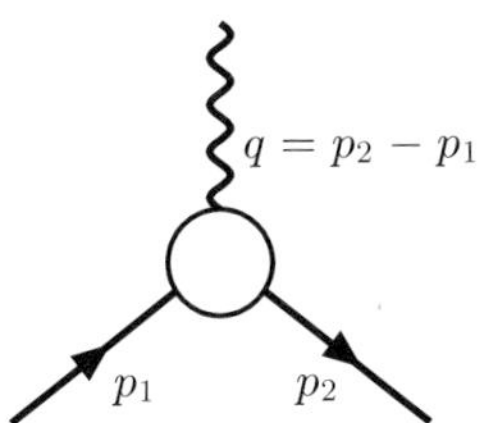

Fig. 2.1. Muon interaction with the external electromagnetic field

K. Melnikov and A. Vainshtein: *Theory of the Muon Anomalous Magnetic Moment*
STMP **216**, 7–32 (2006)
© Springer-Verlag Berlin Heidelberg 2006

where $\sigma^{\mu\nu} = (i/2)(\gamma^\mu\gamma^\nu - \gamma^\nu\gamma^\mu)$. The two form factors, F_D and F_P, are known as the Dirac and Pauli form factors. We need to know the connection of these form factors to the muon anomalous magnetic moment.[1]

In the non-relativistic quantum mechanics, the muon in the electric and magnetic fields is described by the Hamiltonian

$$H = \frac{(\boldsymbol{p} - e\boldsymbol{A})^2}{2m} - \boldsymbol{\mu}\boldsymbol{B} + e\varphi \, , \tag{2.4}$$

where $\boldsymbol{B} = \boldsymbol{\nabla} \times \boldsymbol{A}$ is the external magnetic field and $\boldsymbol{\mu}$ is the muon magnetic moment operator

$$\boldsymbol{\mu} = 2\mu\, \boldsymbol{s} = \mu\,\boldsymbol{\sigma} \, . \tag{2.5}$$

The proportionality coefficient μ in (2.5) is referred to as the *muon magnetic moment*.

To find the relation between the muon magnetic moment μ and the Dirac and Pauli form factors, we consider the scattering of the muon off the external vector potential A_μ in the non-relativistic approximation, using the Hamiltonian (2.4), and compare the result to (2.2).

The non-relativistic scattering amplitude in the Born approximation is given by

$$f = -\frac{m}{2\pi} \int \mathrm{d}^3 r \, \Psi^*(\boldsymbol{p}_2) V \Psi(\boldsymbol{p}_1) \, , \tag{2.6}$$

where $\Psi(\boldsymbol{p}_{1,2}) = \chi_{1,2}\, e^{i\boldsymbol{p}_{1,2}\boldsymbol{r}}$ is the muon wave function in the non-relativistic approximation including spin degrees of freedom described by the Pauli spinor χ and

$$V = -\frac{e}{2m}\,(\boldsymbol{p}\boldsymbol{A} + \boldsymbol{A}\boldsymbol{p}) - \mu\,\boldsymbol{\sigma}\boldsymbol{B} + e\varphi \, . \tag{2.7}$$

Computing f in (2.6) reduces to taking the Fourier transform; we obtain

$$f = -\frac{m}{2\pi}\, \chi_2^+ \left(-\frac{e}{2m}\boldsymbol{A}_q\,(\boldsymbol{p}_2 + \boldsymbol{p}_1) + e\varphi_q - i\mu\,\boldsymbol{\sigma}[\boldsymbol{q} \times \boldsymbol{A}_q] \right) \chi_1 \, , \tag{2.8}$$

where φ_q and $\boldsymbol{A}_q$ stand for the Fourier components of the electric potential φ and the vector potential $\boldsymbol{A}$ and $\boldsymbol{q} = \boldsymbol{p}_2 - \boldsymbol{p}_1$.

We should be able to derive (2.8) starting from the relativistic expression for the scattering amplitude (2.2) and taking the non-relativistic limit. If the Dirac spinors are normalized to $2m$, the relation between the two scattering amplitudes in the non-relativistic limit is [1]

$$-\,i \lim_{|\boldsymbol{p}|\ll m} \mathcal{M} = 4\pi f \, . \tag{2.9}$$

To derive the non-relativistic limit of the amplitude $\mathcal{M}$, (2.2), we use the explicit representation of the Dirac matrices

---

[1] We emphasize that the parametrization of the matrix element (2.3) is the most general one if an only if the external fermions are on the mass-shell $p_1^2 = p_2^2 = m^2$.

$$\gamma^0 = \begin{pmatrix} I & 0 \\ 0 & -I \end{pmatrix}, \qquad \gamma^i = \begin{pmatrix} 0 & \sigma_i \\ -\sigma_i & 0 \end{pmatrix}, \tag{2.10}$$

and the Dirac spinors

$$u_p = \sqrt{E+m} \begin{pmatrix} \chi \\ \dfrac{\boldsymbol{p\sigma}}{E+m}\chi \end{pmatrix}, \tag{2.11}$$

where $E = \sqrt{\boldsymbol{p}^2 + m^2}$. Using these expressions in (2.2) and working through first order in $|\boldsymbol{p}_{1,2}|/m$, we obtain

$$\mathcal{M} = -2iem\chi_2^+ \left[F_D(0)\left(\varphi_q - \frac{\mathbf{A}_q\,(\boldsymbol{p}_2 + \boldsymbol{p}_1)}{2m} \right) - i\,\frac{F_D(0) + F_P(0)}{2m}\,\boldsymbol{\sigma}\,[\boldsymbol{q} \times \mathbf{A}_q] \right] \chi_1 . \tag{2.12}$$

Using (2.9, 2.8) and (2.12) we find

$$F_D(0) = 1 , \qquad \mu = \frac{e}{2m}\,(F_D(0) + F_P(0)) . \tag{2.13}$$

Comparing (1.2, 2.5, 2.13), we find the gyromagnetic factor of the muon

$$g = 2(1 + F_P(0)) . \tag{2.14}$$

Hence, if the Pauli form factor $F_P(q)$ does not vanish for $q = 0$, the gyromagnetic ratio of the muon is different from two, the value predicted by the Dirac equation. It is conventional to call this difference *the muon anomalous magnetic moment* and write

$$\frac{g-2}{2} = a_\mu = F_P(0) . \tag{2.15}$$

In QED, a_μ can be computed in the perturbative expansion in the fine structure constant $\alpha = e^2/(4\pi)$,

$$a_\mu = \sum_{i=1}^{\infty} a_\mu^{(i)} = \sum_{i=1}^{\infty} c_i \left(\frac{\alpha}{\pi} \right)^i . \tag{2.16}$$

The first term in the series is $\mathcal{O}(\alpha)$ since, when radiative corrections are neglected, the Pauli form factor vanishes. This is easily seen from the QED Lagrangian (2.1) which implies that, through leading order in α, the interaction between the external electromagnetic field and the muon is given by $-ieu_{p_2}\gamma^\mu u_{p_1} A_\mu$.

In QED, a naive application of the perturbative expansion leads to a confusing situation since results often appear to be divergent. Such divergences are removed by a procedure known as the "renormalization". The idea is to view all the entries in the QED Lagrangian (2.1) as "bare" quantities, whereas all physically meaningful results are expressed through "renormalized" quantities. The renormalization is multiplicative, e.g. $e = Z_e e_{\mathrm{phys}}$, $m = Z_m m_{\mathrm{phys}}$, $\psi = Z^{1/2} \psi_{\mathrm{phys}}$, etc. The renormalization constants are fixed by requiring that Green's functions, computed with the QED Lagrangian, satisfy certain conditions. For example, the mass renormalization constant Z_m is determined from the requirement that m_{phys} is the physical mass of the muon and therefore the muon propagator should have a pole at $p^2 = m_{\mathrm{phys}}^2$. Similarly, the charge renormalization constant Z_e is determined by requiring that e_{phys} gives the strength of the interaction of an on-shell photon with a fermion. A consequence of the current conservation is the fact that the Dirac form factor satisfies the condition $F_D(0) = 1$ to *all orders in the perturbative expansion*. The renormalization constants influence the Pauli form factor only indirectly, through the mass, charge and the fermion wave function renormalization, because there is no corresponding tree-level operator in the QED Lagrangian. Therefore, the anomalous magnetic moment is the unique prediction of QED; moreover, the $\mathcal{O}(\alpha)$ contribution to a_μ has to be finite without any renormalization.

The QED radiative corrections provide the largest contribution to the muon anomalous magnetic moment. The one-loop result was computed by Schwinger in 1949 [2]; currently, QED calculations have been extended to the four-loop order and even some estimates of the five-loop contribution exist. It is interesting to remark that Schwinger's calculation was performed *before* the renormalizability of QED was understood in detail. Historically, this provides an interesting example of a fundamental physics result derived from a theory that was considered ambiguous at that time [3].

2.2 Features of the Radiative Corrections to the Muon Magnetic Anomaly

Before describing the computation of the muon anomalous magnetic moment in QED, we discuss general features of the radiative corrections to this quantity. When QED, electroweak or hadronic corrections to a_μ are considered, the anomalous magnetic moment receives contributions from all the existing particles. It is therefore useful to understand how the contribution of a particle with the mass m_p to the anomalous magnetic moment depends on its mass. First, consider the case $m_p \geq m$. Then, in general,

$$\delta a_\mu^p \sim \left(\frac{\alpha}{\pi}\right)^{n_p} \frac{m^2}{m_p^2} \ln^{k_p} \frac{m_p}{m} \, , \tag{2.17}$$

where n_p, k_p depend on the order of the perturbative expansion in the fine structure constant at which this contribution appears. The power of the logarithm k_p can not be determined on general grounds.

To understand the m^2/m_p^2 suppression factor in (2.17), we note that when a contribution of a particle with the mass $m_p \gg m$ to the muon anomalous magnetic moment is considered, the loop momenta $k \sim m_p$ provide the dominant contribution. In this situation it is natural to consider the limit of the vanishing muon mass m. QED interactions conserve the muon helicity in that limit. Note, however, that the term $F_D \sigma_{\mu\nu} q^\nu$ in the scattering amplitude (2.3) changes the muon helicity, whereas the term proportional to the Dirac form factor conserves the helicity. It follows that at $m = 0$, loops with heavy particles contribute to the Dirac, but not to the Pauli form factor. Corrections to the Pauli form factor and, hence, to the muon anomalous magnetic moment, require helicity flip; this can be achieved by a single mass insertion into one of the muon lines. Using the fact that the vertex function $\Gamma_\mu(p_1, p_2)$ in (2.3) is dimensionless, we arrive at

$$\Gamma_\mu \overset{a_\mu}{\sim} \left(\frac{\alpha}{\pi}\right)^{n_p} \frac{m\,\sigma_{\mu\nu} q^\mu}{m_p^2} \,, \tag{2.18}$$

from where (2.17) immediately follows.

If $m_p \ll m$, the correction to a_μ can be estimated as

$$\delta a_\mu^p \sim \left(\frac{\alpha}{\pi}\right)^{n_p} \ln^{k_p} \frac{m}{m_p} \,, \tag{2.19}$$

where the power of the logarithm k_p satisfies $k_p < n_p$ but, otherwise, can not be determined without detailed analysis. Specific examples that lead to (2.19) are discussed in the following sections.

Equations (2.17, 2.19) are useful for understanding a very different structure of higher order corrections to the electron and the muon anomalous magnetic moments. Since electron is the lightest charged particle, the corrections to the electron anomalous magnetic moment are governed by powers of α/π; contributions due to heavier particles are all power-suppressed by at least $m_e^2/m^2 \sim 10^{-4}$. Hadronic contributions that are mostly determined by the mass of the ρ meson and are responsible for a large fraction of the uncertainty in the theoretical prediction for the muon anomalous magnetic moment, are tiny for the electron magnetic anomaly because of m_e^2/m_ρ^2 suppression. On the contrary, in case of the muon anomalous magnetic moment the expansion parameter in QED is $(\alpha/\pi)\ln(m/m_e)$ and, also, the hadronic corrections are significant since the mass of the ρ meson is not too different from the mass of the muon.

There are two interesting consequences of these considerations. First, since both hadronic and potential new physics effects on the electron anomalous magnetic moment are strongly suppressed, the measured value of the electron anomalous magnetic moment can be used to determine the value of the fine structure constant α with high precision [4].

Second, the beyond the Standard Model physics should contribute much stronger to the muon anomalous magnetic moment, making this observable an attractive laboratory for low-energy tests of the Standard Model. This feature was recognized already in late fifties [5] when it was pointed out that if QED is modified at some energy scale Λ, there is a correction to the anomalous magnetic moment of the muon of the order of m^2/Λ^2. At that time, the goal was to test QED up to the energies around 4 GeV; currently, the goal is to detect contributions of particles with masses of the order of *a few hundred* GeV through the measurement of the muon magnetic anomaly. This illustrates a tremendous progress that the high energy physics made in the past half a century.

2.3 One-loop QED Contribution: Schwinger Correction

The Schwinger correction can be obtained by direct evaluation of the Feynman diagram shown in Fig. 2.2. The expression for $\Gamma_\mu(p_2, p_1)$ in the Feynman gauge reads

$$\Gamma_\mu(p_2, p_1) = -ie^2 \int \frac{\mathrm{d}^4 k}{(2\pi)^4} \frac{\gamma_\alpha(\hat{k} + \hat{p}_2 + m)\gamma_\mu(\hat{k} + \hat{p}_1 + m)\gamma^\alpha}{k^2[(k+p_2)^2 - m^2][(k+p_1)^2 - m^2]} , \qquad (2.20)$$

where we use $\hat{k} = \gamma^\alpha k_\alpha$. In principle, (2.20) is divergent both in the ultra-violet $k \to \infty$ and in the infra-red $k \to 0$. However, it is easy to see that those divergences affect the Dirac form factor only.

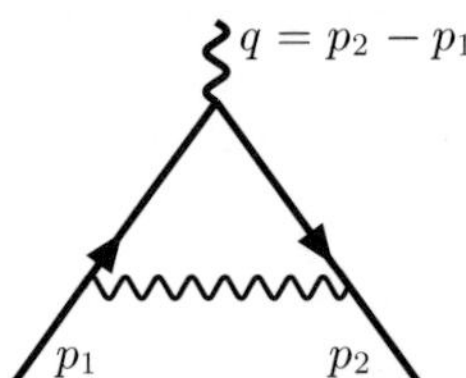

Fig. 2.2. A one-loop QED diagram that leads to the muon anomalous magnetic moment

Consider first the ultra-violet divergence; since it originates from large values of k, the dependence on p_1 and p_2 in (2.20) can be neglected. We obtain

$$\Gamma_\mu(p_2, p_1) \xrightarrow{\mathrm{UV}} -ie^2 \int \frac{\mathrm{d}^4 k}{(2\pi)^4} \frac{\gamma_\alpha \hat{k}\gamma_\mu\hat{k}\gamma^\alpha}{k^6} = -ie^2\gamma_\mu \int \frac{\mathrm{d}^4 k}{(2\pi)^4 k^4} , \qquad (2.21)$$

which implies that the ultra-violet divergence is only present in the Dirac form factor.

The infra-red divergence originates from small virtual momenta $k \to 0$; in that case, we can neglect the dependence of the numerator in (2.20) on k. We then use the fact that the vertex function Γ_μ is needed in the combination with two on-shell spinors $\bar{u}_{p_2}\Gamma_\mu u_{p_1}$. This allows us to write

$$\Gamma_\mu(p_2, p_1) \xrightarrow{\text{IR}} -ie^2 \int \frac{\mathrm{d}^4 k}{(2\pi)^4} \frac{\gamma_\alpha(\hat{p}_2 + m)\gamma_\mu(\hat{p}_1 + m)\gamma^\alpha}{k^2(2p_2 k)(2p_1 k)}$$

$$= -ie^2\gamma_\mu \int \frac{\mathrm{d}^4 k}{(2\pi)^4} \frac{p_2 p_1}{k^2(p_2 k)(p_1 k)} \, , \tag{2.22}$$

and proves that the infra-red divergence occurs in the Dirac form factor only.

As the consequence of the fact that the divergencies do not affect the Pauli form factor, we may compute the anomalous magnetic moment without any regulator. Starting from (2.20), we combine the denominators using Feynman parameters, integrate over the loop momentum and drop all the terms that are proportional to γ_μ and hence do not contribute to the Pauli form factor. We obtain

$$\Gamma_\mu(p_1, p_2) \stackrel{a_\mu}{=} \frac{\alpha}{2\pi} \int [\mathrm{d}x]_3 \frac{\hat{\mathcal{P}}\gamma_\mu\hat{\mathcal{P}} + \hat{p}_1\hat{\mathcal{P}}\gamma_\mu + \gamma_\mu\hat{\mathcal{P}}\hat{p}_2}{\mathcal{P}^2} \, , \tag{2.23}$$

where $[\mathrm{d}x]_3 = \mathrm{d}x_1 \mathrm{d}x_2 \mathrm{d}x_3 \delta(1 - x_1 - x_2 - x_3)$ and $\mathcal{P} = p_1 x_1 + p_2 x_2$.

To derive a_μ, we only have to compute (2.23) through *linear* terms in the momentum transfer q. Then, the dependence on q^2 in $\mathcal{P}^2$ can be neglected

$$\mathcal{P}^2 = m^2(x_1 + x_2)^2 - q^2 x_1 x_2 \to m^2(x_1 + x_2)^2. \tag{2.24}$$

We can also simplify the numerator in (2.23) by systematically projecting out the Dirac structure γ_μ and keeping q through first order. Then, for example,

$$\hat{p}_1\hat{\mathcal{P}}\gamma_\mu = \left(m^2 x_1 + x_2\hat{p}_1\hat{p}_2\right)\gamma_\mu \to x_2\hat{p}_1\hat{p}_2\gamma_\mu \to -x_2 m\hat{p}_1\gamma_\mu \to x_2 m\hat{q}\gamma_\mu \, .$$

Performing similar manipulations with other terms in the numerator of (2.23), we obtain the following expression for the part of Γ_μ that gives rise to the muon anomalous magnetic moment

$$\Gamma_\mu \stackrel{a_\mu}{=} \frac{\alpha}{2\pi m} \int [\mathrm{d}x]_3 \frac{\hat{q}\gamma_\mu(x_2(1 - x_1) - x_1^2) - \gamma_\mu\hat{q}(x_1(1 - x_2) - x_2^2)}{(x_1 + x_2)^2} \, . \tag{2.25}$$

The integration in (2.25) is easily performed with the change of variables $x_1 = yz$ and $x_2 = y(1 - z)$; this leads to $x_1 + x_2 = y$ and $[\mathrm{d}x]_3 = y\mathrm{d}y\mathrm{d}z$, with integrations over y, z from 0 to 1. We then obtain

$$\Gamma_\mu \stackrel{a_\mu}{=} \frac{\alpha}{2\pi} \frac{i\sigma_{\mu\nu}q^\nu}{2m} \, . \tag{2.26}$$

Comparing (2.26) to (2.3), we find

$$a_\mu^{(1)} = F_P(0) = \frac{\alpha}{2\pi} \, . \tag{2.27}$$

Equation (2.27) is the celebrated Schwinger correction [2]. Historically, Schwinger derived this result for the electron, rather than the muon, to explain the measurement of the hyperfine splittings in gallium atoms by Foley and Kusch [6]. However, as it is seen from (2.27), the one-loop QED contribution to the anomalous magnetic moment does not depend on the mass of the fermion and is, therefore, the same for muons and electrons.

Although Feynman parameters can be used efficiently in one-loop calculations, they are not very practical for computations in higher orders. The technique that is used successfully for cutting edge calculations in perturbative QED is based on the integration-by-parts identities for Feynman diagrams [7], developed for applications in perturbative Quantum Chromodynamics. Below, we introduce this technique by re-calculating the one-loop QED contribution to a_μ.

As we said earlier, the anomalous magnetic moment does not require the renormalization in QED; once perturbative calculations are performed and the result is expressed through the physical charge and mass of the muon, the anomalous magnetic moment is finite. However, for practical computation of QED corrections in higher orders, it is convenient to introduce both an ultra-violet and an infra-red regularizations. This is required because in higher orders of perturbation theory, where many Feynman diagrams contribute, individual Feynman graphs become divergent and the finite result for a_μ is obtained only after contributions of individual Feynman graphs in a given order of perturbation theory are summed up. For practical reasons, it is more convenient to deal with one diagram at a time; this necessitates introducing the regularization at intermediate steps.

Although there are different ways to regularize ultra-violet and infra-red divergences in QED, it is convenient to choose dimensional regularization [8, 9, 10], analytically continuing the space-time dimensionality to $d = 4 - 2\epsilon$. A consistent implementation of the dimensional regularization leads to a set of rules for Feynman integrals [11] that imply that those integrals, in general, can not be understood as Lebesque integrals; however, the new definition of the integrals enables simultaneous regularization of both ultra-violet and infra-red divergencies. As a consequence, the dimensional regularization is very economical from the computational viewpoint but requires care when a physical interpretation of divergent expressions is attempted.

We consider the one-loop vertex function (2.20) and continue it to d dimensions; this leads to the modification of the integration measure $\mathrm{d}^4k/(2\pi)^4 \to \mathrm{d}^d k/(2\pi)^d$. If the integration over the loop momentum k is performed, the result contains both the Dirac and the Pauli form factors, (2.3). Since we are not interested in the Dirac form factor, this is inconvenient, especially when higher order corrections to the anomalous magnetic moment are considered. The standard way to deal with this issue is to construct an operator that projects the vertex function $\Gamma_\mu(p_2, p_1)$ on to the Pauli form factor. Such operator reads

$$\Pi_\mu = \frac{2m^2(\hat{p}_1 + m)}{(d-2)q^2(4m^2 - q^2)} \left(\gamma_\mu + \frac{((4-2d)q^2 - 8m^2)}{(4m^2 - q^2)} \frac{P_\mu}{4m} \right) (\hat{p}_2 + m) ,$$

$$(2.28)$$

where $q = p_2 - p_1$ and $P = p_2 + p_1$. The Pauli form factor is obtained if the trace of $\Gamma_\mu(p_2, p_1)$ with Π_μ is computed and the sum over Lorentz index μ is performed

$$\mathrm{Tr}\left[\Pi^\mu \Gamma_\mu(p_2, p_1)\right] = F_P(q^2) , \tag{2.29}$$

Because the anomalous magnetic moment is given by $F_P(0)$, the left hand side of (2.29) has to be evaluated at $q^2 = 0$. Here, care should be exercised since there is an explicit $1/q^2$ factor in the expression for Π_μ, (2.28). To elucidate how this "singularity" disappears, consider the limit of (2.28, 2.29) for small values of q; the approximate expression for Π_μ reads

$$\Pi_\mu \stackrel{q \ll m}{\approx} \frac{t_{\mu\nu}}{q^2}\, q^\nu , \qquad t_{\mu\nu} = -\frac{i(\hat{p}_1 + m)\sigma_{\mu\nu}\,(\hat{p}_2 + m)}{4m(d-2)} . \tag{2.30}$$

It follows from this asymptotics that when (2.29) is computed for a Feynman diagram that contributes to the muon anomalous magnetic moment and the limit of small q is taken, the result is finite because of the on-shell conditions for the external muon that imply $p_1 q = -q^2/2, \quad p_2 q = q^2/2$. Hence, each Feynman diagram provides a non-singular contribution to $F_P(q^2)$, in the limit $q^2 \to 0$.

To expand (2.29) in powers of q^2 for the one-loop Feynman diagram (2.20), we compute the trace and substitute $p_2 = p_1 + q$, both in the numerator and in the appropriate propagators. Expanding the propagators in powers of q, we arrive at the expression that can symbolically be written as

$$F_P(q^2) \sim \frac{1}{q^2} \int \frac{\mathrm{d}^d k}{(2\pi)^d} \sum_{n=2}^{\infty} \frac{\mathrm{Num}_n(q^2, kq, p_1 q)}{k^2[(k+p_1)^2 - m^2]^n} , \tag{2.31}$$

where the numerator scales as first power of q, for $q \to 0$. We remove the scalar product $p_1 q$ using the on-shell condition.

To remove the scalar product kq, we write

$$q = -\frac{q^2}{2m^2}\, p_1 + q_\perp , \qquad p_1 q_\perp = 0 , \tag{2.32}$$

and observe that the left hand side of (2.31) is a function of $q_\perp^2$ only. We therefore can average (2.31) over the directions of the vector $q_\perp$. The averaging is performed easily with the help of the following formula

$$(kq_\perp)^n \to \begin{cases} (k_\perp^2 q_\perp^2)^{n/2}\, \dfrac{n!}{2^n (n/2)!}\, \dfrac{\Gamma((d-1)/2)}{\Gamma((d-1)/2 + n/2)} , & n \text{ even} , \\[2ex] 0 , & n \text{ odd} , \end{cases} \tag{2.33}$$

where $k_\perp^2 = k^2 - (kp_1)^2/m^2$ and $q_\perp^2 = q^2(1 - q^2/(4m^2))$. After the averaging, the right hand side of (2.31) becomes a function of q^2; the factor $1/q^2$ cancels out and, taking the limit $q^2 \to 0$, we arrive at the expression for $F_P(0)$. It is easy to see that $F_P(0)$ can be expressed through a linear combination of integrals

$$S_M(a_1, a_2) = \int \frac{\mathrm{d}^d k}{\pi^{d/2}} \frac{1}{(k^2)^{a_1}(k^2 + 2kp)^{a_2}} \,, \tag{2.34}$$

with $p^2 = m^2$. It is convenient to perform the Wick rotation and deal with the integrals

$$S_1(a_1, a_2) = \int \frac{\mathrm{d}^d k}{\pi^{d/2}} \frac{1}{(k^2)^{a_1}(k^2 + 2kp)^{a_2}} \,, \tag{2.35}$$

defined in Euclidean space with $p^2 = -m^2$.

Those integrals can be separated into three different classes. The first class consists of integrals $S_1(a_1, a_2)$ with the second argument $a_2 \leq 0$. It is easy to see that all such integrals vanish since, in dimensional regularization,

$$\int \mathrm{d}^d k (k^2)^a = 0 \,, \tag{2.36}$$

for any a .

The second class consists of integrals $S_1(a_1, a_2)$ with $a_1 \leq 0$. For these integrals, we shift the integration momenta $k \to k - p$. Upon the shift,

$$S_1(a_1 \leq 0, a_2) = \int \frac{\mathrm{d}^d k}{\pi^{d/2}} \frac{(k^2 - 2pk - m^2)^{-a_1}}{(k^2 + m^2)^{a_2}} \,. \tag{2.37}$$

We now expand the numerator, average over directions of the vector k and use $k^2 = (k^2 + m^2) - m^2$, to remove k^2 from the numerator. Upon doing so, we see that all the integrals $S_1(a_1 \leq 0, a_2)$ can be expressed as linear combinations of the integrals $S_1(0, a_2)$ with various values of the second argument a_2. We will discuss computation of those integrals shortly.

The last class we have to consider consists of integrals $S_1(a_1 \geq 0, a_2 \geq 0)$. Such integrals are conveniently studied with the help of the integration-by-parts identities [7]. The derivation of the integration-by-parts identities is based on the observation that, in dimensional regularization, the integral of the total derivative vanishes. Therefore,

$$0 = \int \frac{\mathrm{d}^d k}{\pi^{d/2}} \frac{\partial}{\partial k_\mu} \frac{l_\mu}{(k^2)^{a_1}(k^2 + 2pk)^{a_2}} \,, \tag{2.38}$$

for any vector l. It is convenient to take $l = k$. Then

$$(d - 2a_1 - a_2)S_1(a_1, a_2) - a_2 S_1(a_1 - 1, a_2 + 1) = 0 \,. \tag{2.39}$$

This equation relates two integrals with different values of a_1. For $a_1 > 0$, we can solve (2.39) for the first term to reduce a_1 by one unit. Repeating this operation, we can express all the integrals $S_1(a_1 \geq 0, a_2 \geq 0)$ through a set of integrals $S_1(0, a_2)$.

We are thus left with the integrals $S_1(0, a_2)$. For those integrals, we shift $k \to k - p$, removing the dependence on the momentum p, and write

$$S_1(0, a_2) = \int \frac{\mathrm{d}^d k}{(\pi)^{d/2}} \frac{1}{(k^2 + m^2)^{a_2}} \; . \tag{2.40}$$

Constructing the integration-by-parts equation for these integrals we obtain

$$(d - 2a_2)S_1(0, a_2) + 2a_2 m^2 S_1(0, a_2 + 1) = 0 \; . \tag{2.41}$$

By solving for the second term on the left hand side of this equation, we obtain the recurrence relation that allows us to reduce the argument a_2 by one unit. Repeated application of this recurrence relation maps any integral $S_1(0, a_2)$ on to $S_1(0, 1)$.

As a consequence, we can express *any* integral $S_1(a_1, a_2)$, that appears in the expression for a_μ, through $S_1(0, 1)$ in an *entirely algebraic way*. For the anomalous magnetic moment, we derive

$$a_\mu^{(1)} = -\frac{\alpha}{\pi} \frac{(d - 5)(d - 2)(d - 4)}{8(d - 3)} \frac{S_1(0, 1)}{(4\pi)^{(d-4)/2} m^2} \; . \tag{2.42}$$

The integral $S_1(0, 1)$ is easy to compute by integrating over the loop momentum k; the result reads

$$S_1(0, 1) = m^{d-2}\Gamma(1 - d/2) = -m^{2-2\epsilon} \frac{\Gamma(1 + \epsilon)}{\epsilon(1 - \epsilon)} \; . \tag{2.43}$$

Using (2.42, 2.43) we recover the muon anomalous magnetic moment through first order in the perturbative expansion in QED in (2.27).

Alternatively, we could have expressed a_μ through the integral $S_1(0, 3)$ which is *finite* for $d = 4$, $S_1(0, 3) = \Gamma(1 + \epsilon)/(2m^{2+2\epsilon})$. We then obtain

$$a_\mu^{(1)} = -\frac{\alpha}{\pi} \frac{(d - 5)}{(d - 3)} \frac{m^2 S_1(0, 3)}{(4\pi)^{(d-2)/2}} = \frac{\alpha}{2\pi} \; . \tag{2.44}$$

2.4 Two-loop QED Corrections to a_μ

In this section we discuss the two-loop QED corrections to the muon anomalous magnetic moment. It is much harder to compute those corrections than the one-loop ones. In addition, at this order in the perturbative expansion a new qualitative feature appears – the anomalous magnetic moment becomes dependent on the masses of all charged particles through vacuum polarization diagrams. It is conventional to include electrons, muons and τ leptons when discussing the QED radiative corrections; the contribution of other particles, e.g. quarks and electroweak gauge bosons, is relegated to hadronic and electroweak corrections, respectively. We write the $\mathcal{O}(\alpha^2)$ contribution to a_μ as the sum of three terms

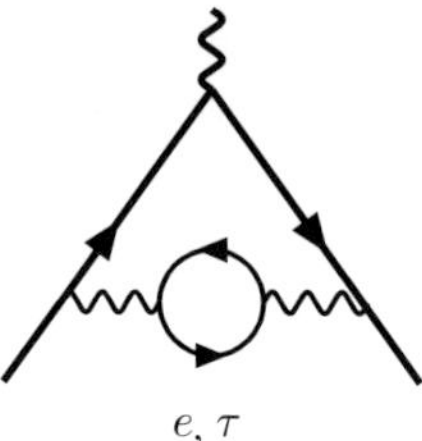

Fig. 2.3. Vacuum polarization contribution to the muon anomalous magnetic moment due to electron and tau loops

$$a_\mu^{(2)} = \left(\frac{\alpha}{\pi}\right)^2 \left[w_2^{\mathrm{vp},e} + w_2^{\mathrm{vp},\tau} + w_2\right] . \tag{2.45}$$

Here, $w_2^{\mathrm{vp},e(\tau)}$ refers to the vacuum polarization diagrams due to electron and τ loops, Fig. 2.3; the term w_2 contains the contributions of all the diagrams where only muons appear, Fig. 2.4. The two-loop QED corrections to a_μ were computed in [12, 13, 14], approximately half a century ago.

We begin the discussion of the two-loop corrections to the muon anomalous magnetic moment with the vacuum polarization diagrams Fig. 2.3 where electron and τ loops are considered. The muon vacuum polarization contribution will be treated later, alongside with other diagrams Fig. 2.4 that only depend on the muon mass. Because of the mass hierarchy, $m_e \ll m \ll m_\tau$, the electron and τ contributions to $a_\mu^{(2)}$ are quite different, as we explain below.

The vacuum polarization contribution to a_μ is computed by applying the projection operator (2.28) to the vertex function $\Gamma_\mu(p_2, p_1)$ corresponding to the Feynman diagram Fig. 2.3. The vertex function can be written as

$$\Gamma_\mu = -e^4 \int \frac{\mathrm{d}^4 k}{(2\pi)^4} \, \frac{\bar{u}(p_2)\gamma_\alpha(\hat{p}_2 + \hat{k} + m)\gamma_\mu(\hat{p}_1 + \hat{k} + m)\gamma_\beta u(p_1)}{k^4((p_2 + k)^2 - m^2)((p_1 + k)^2 - m^2)} \, \Pi^{\alpha\beta}(k) , \tag{2.46}$$

where

$$\Pi^{\alpha\beta}(k) = -i\left(g^{\alpha\beta}k^2 - k^\alpha k^\beta\right)\Pi(k^2) ,$$

is the general form of the photon self-energy consistent with the electromagnetic gauge invariance. The photon self-energy satisfies the subtracted dispersion relation

$$\Pi(k^2) = \frac{k^2}{\pi} \int\limits_{s_0}^{\infty} \frac{\mathrm{d}s}{s(s - k^2 - i0)} \, \mathrm{Im}\Pi(s) , \tag{2.47}$$

where s_0 refers to the lowest invariant mass squared of the fermion pair that can be produced in a decay of a virtual photon with the invariant mass k^2. For the diagram Fig. 2.3, $s_0 = 4m_e^2$ or $4m_\tau^2$.

Using the optical theorem [1], the imaginary part of the photon self-energy can be expressed through the e^+e^- annihilation cross-section into a lepton pair

$$\mathrm{Im}\Pi(s) = \frac{R^{\mathrm{lept}}(s)}{12\pi} \,, \qquad R^{\mathrm{lept}}(s) = \frac{\sigma_{\mathrm{lept}}(s)}{\sigma_{\mathrm{point}}(s)} \,, \tag{2.48}$$

where

$$\sigma_{\mathrm{lept}} = \sigma_{\mathrm{point}} \sqrt{1 - \frac{4m_{\mathrm{lept}}^2}{s}} \left(1 + \frac{2m_{\mathrm{lept}}^2}{s}\right), \qquad \sigma_{\mathrm{point}} = \frac{4\pi\alpha^2}{3s} \,. \tag{2.49}$$

We insert the photon self-energy (2.47) into (2.46) and integrate over the loop momentum k. This is easily done, since the factor $(-g_{\mu\nu}+k_\mu k_\nu/k^2)/(k^2 - s + i0)$ introduced to the integrand (2.46) by the dispersion representation (2.47) is the propagator of a gauge boson with the mass $\sqrt{s}$. Hence, the integration over the loop momentum is standard and can be easily performed by introducing e.g. the Feynman parameters. Integrating over the loop momentum, we arrive at the following representation for the hadronic vacuum polarization contribution to a_μ

$$w_2^{\mathrm{vp,lept}} = \frac{\pi}{3\alpha} \int_{4m_{\mathrm{lept}}^2}^{\infty} \frac{ds}{s} \, R^{\mathrm{lept}}(s) \, a_\mu^{(1)}(s) \,. \tag{2.50}$$

Here, $a_\mu^{(1)}(s)$ refers to the one-loop contribution to the muon anomalous magnetic moment of a neutral vector boson with the mass $\sqrt{s}$ and the electromagnetic coupling to the muon. The corresponding expression reads

$$a_\mu^{(1)}(s) = \left(\frac{\alpha}{\pi}\right) \left[\frac{1}{2} - \rho + \frac{\rho(\rho - 2)\ln(\rho)}{2} - \left(1 - 2\rho + \frac{\rho^2}{2}\right) f(s)\right] \,, \tag{2.51}$$

where

$$f(s) = \left[\theta(s - 4m^2) \frac{1}{\beta} \ln \frac{1+\beta}{1-\beta} + \theta(4m^2 - s) \frac{2\arctan(\tilde{\beta})}{\tilde{\beta}}\right] \,, \tag{2.52}$$

with $\rho = s/m^2$, $\beta = \sqrt{1 - 4/\rho}$, $\tilde{\beta} = \sqrt{4/\rho - 1}$. In the limits of small and large values of s/m^2, (2.51) reads

$$a_\mu^{(1)}(s) \approx \frac{\alpha}{\pi} \times \begin{cases} \dfrac{m^2}{3s}\,, & s \gg m^2; \\[2ex] \dfrac{1}{2} - \dfrac{\pi\sqrt{s}}{2m}\,, & s \ll m^2 \,. \end{cases} \tag{2.53}$$

Using the explicit form of the function $a_\mu^{(1)}(s)$, the integration over s in (2.50) can be performed analytically [15]. The result of [15] in a simplified form [16] reads

$$w_2^{\text{vp,lept}} = -\frac{25}{36} - \frac{\ln r}{3} + r^2(4 + 3\ln r) + r^4\left(\frac{\pi^2}{6} + 2\ln^2 r + \text{Li}_2(1 - r^2)\right)$$

$$+ \frac{r}{2}\left(1 - 5r^2\right)\left[\frac{\pi^2}{3} + \ln r \ln(1 + r) + \text{Li}_2(1 - r) + \text{Li}_2(-r)\right] , \quad (2.54)$$

where $r = m_{\text{lept}}/m$ and $\text{Li}_2(r) = -\int_0^r (dt/t)\ln(1 - t)$ is the dilogarithmic function.

Knowledge of the exact result (2.54) is important for deriving the Standard Model prediction for a_μ. However, the limiting cases $m_{\text{lept}} \ll m$ and $m_{\text{lept}} \gg m$ can be easily obtained from the integral representation (2.50); such a derivation is instructive since it offers useful insights into the underlying physics.

First, consider the case $m_{\text{lept}} = m_e \ll m$. It is easy to see that $w_2^{\text{vp},e}$ contains a logarithmic enhancement factor $\ln(m/m_e)$ which can only appear from the integration over s in (2.50) if $m_e^2 \ll s \ll m^2$. In that limit $a_\mu^{(1)}(s)$ can be approximated by $a_\mu^{(1)}(0) = \alpha/(2\pi)$ and $R(s)$ can be approximated by its value at $s = \infty$, $R(s) = 1$. This gives

$$w_2^{\text{vp},e} \approx \frac{1}{6}\int_{4m_e^2}^{4m^2} \frac{ds}{s} = \frac{1}{3}\ln\frac{m}{m_e} . \quad (2.55)$$

We could have guessed this result without any calculation. Indeed, in QED, as in any other field theory, the coupling constant depends on the momentum scale and satisfies the renormalization group equation [17]

$$\mu\frac{d}{d\mu}\left(\frac{\alpha(\mu)}{\pi}\right) = \frac{1}{\pi}\beta(\alpha) = \frac{2}{3}\left(\frac{\alpha(\mu)}{\pi}\right)^2 + \mathcal{O}(\alpha^3) . \quad (2.56)$$

Through $\mathcal{O}(\alpha^2)$, the solution to this equation yields

$$\frac{\alpha(\mu)}{\alpha} = 1 + \frac{2\alpha}{3\pi}\ln\frac{\mu}{m_e} , \quad (2.57)$$

where $\alpha = \alpha(m_e)$ is the conventional fine-structure constant $\alpha \approx 1/137.06$. When studying a physical process, it is beneficial to choose the parameter μ to be of the order of energy or momentum that are characteristic for the process. For the muon anomalous magnetic moment, the typical virtuality of the photon in the one-loop diagram Fig. 2.1 is clearly of order m^2. To avoid large logarithms in the perturbative expansion for a_μ, we should have written

$$a_\mu^{(1)} = \frac{\alpha(m)}{2\pi} , \quad (2.58)$$

for the one-loop result. Expanding this equation to second order in α using (2.57) for $\mu = m$ and comparing the result with (2.55), we find the complete agreement, as expected.

It is instructive to consider further terms in the expansion of $w_2^{\mathrm{vp,e}}$ in the ratio m_e/m. Expanding (2.54) around $r = 0$, we find

$$w_2^{\mathrm{vp,e}} = \frac{1}{3}\ln\frac{m}{m_e} - \frac{25}{36} + \frac{\pi^2}{4}\frac{m_e}{m} + \mathcal{O}\left(\frac{m_e^2}{m^2}\right). \tag{2.59}$$

An interesting term in this equation is the linear term in the mass ratio since it is non-analytic in the mass of the electron squared m_e^2 and, therefore, can only appear from a peculiar momentum configuration. We may approach the computation of the $\mathcal{O}(m_e/m)$ term in (2.59) starting from the convolution integral (2.50). Then, we require $a_\mu^{(1)}(s)$ in the limit of small s beyond the leading term $a_\mu^{(1)}(0)$. In particular, we are interested in all terms that behave like $\sqrt{s}/m$, for $s \ll m^2$. It is easy to see from (2.50) that if such terms appear, the integration over s produces m_e/m contribution to $w_2^{\mathrm{vp,e}}$.

We now describe a calculation of the $\sqrt{s}/m$ correction to $a_\mu^{(1)}$. At small values of s, the asymptotics of the anomalous magnetic moment is shown in (2.53); since we are interested in the term $\mathcal{O}(\sqrt{s}/m)$, it is convenient to compute the derivative of the anomalous magnetic moment with respect to s. To this end, consider the derivative of the vertex function

$$\frac{d\Gamma_\mu}{ds} = -ie^2\int\frac{d^4k}{(2\pi)^4}\frac{\gamma^\alpha(\hat{k}+\hat{p}_2+m)\gamma_\mu(\hat{k}+\hat{p}_1+m)\gamma^\beta\left(g_{\alpha\beta}-k_\alpha k_\beta/k^2\right)}{(k^2-s)^2[(k+p_2)^2-m^2][(k+p_1)^2-m^2]}. \tag{2.60}$$

We are interested in such terms in (2.60) that lead to $d\Gamma_\mu/ds \sim 1/\sqrt{s}$. From (2.60) it is easy to see that this is only possible if the loop momenta are small, $k \sim \sqrt{s}$. In that limit the muon propagators become *static*, since $k^2 + 2p_{1,2}k + i0 \approx 2p_{1,2}k + i0 \approx 2m(k_0 + i0)$. Note that this is very similar to the heavy quark limit in physics of B-mesons which is studied using the so-called Heavy Quark Effective Theory [18].

Simplifying the numerator in (2.60) and keeping only such terms that contribute to the muon anomalous magnetic moment, we obtain

$$\frac{d\Gamma_\mu}{ds} \stackrel{a_\mu}{=} \frac{e^2\sigma_{\mu\nu}q^\nu}{2m^2}\int\frac{d^4k}{(2\pi)^4}\frac{1}{(k^2-s)^2(k_0+i0)}. \tag{2.61}$$

Comparing (2.61) with (2.3), we derive

$$\frac{da_\mu^{(1)}(s)}{ds} = -\frac{ie^2}{m}\int\frac{d^4k}{(2\pi)^4}\frac{1}{(k^2-s)^2(k_0+i0)}. \tag{2.62}$$

To integrate over k_0, we write

$$\frac{1}{k_0+i0} = \mathrm{P}\left[\frac{1}{k_0}\right] - i\pi\delta(k_0), \tag{2.63}$$

and note that the principal value term can be dropped in (2.62) since the photon propagator is an even function of k_0. Hence, we obtain

$$\frac{\mathrm{d}a_\mu^{(1)}(s)}{\mathrm{d}s} = -\frac{\alpha}{4\pi^2 m} \int \frac{\mathrm{d}^3 k}{(\boldsymbol{k}^2 + s)^2} = -\left(\frac{\alpha}{\pi}\right)\frac{\pi}{4m\sqrt{s}}. \tag{2.64}$$

On account of the boundary condition $a_\mu^{(1)}(0) = \alpha/(2\pi)$, (2.64) leads to

$$a_\mu^{(1)}(s) \approx \frac{\alpha}{\pi}\left(\frac{1}{2} - \frac{\pi\sqrt{s}}{2m}\right). \tag{2.65}$$

Equation (2.65) implies that for the momentum configuration in the loop integrals where the muon is close to the mass shell, the expansion parameter is α, rather than conventional (α/π). This happens because in that momentum configuration, the muon is the source of the time-independent Coulomb (or Yukawa) potential; (2.63) shows explicitly how additional factor of π is generated.

We may now integrate the last term in (2.65) over s using (2.50). The integral diverges quadratically and, therefore, is ill-defined. However, the dependence of the integral on a *linear* term in the mass of the electron m_e is unambiguous. We obtain

$$\delta w_2^{\mathrm{vp},e} = -\frac{\pi m_e}{3m}\left[B\left(-\frac{1}{2},\frac{3}{2}\right) + \frac{1}{2}B\left(\frac{1}{2},\frac{3}{2}\right)\right] = \frac{\pi^2 m_e}{4m}, \tag{2.66}$$

that coincides with the linear term in the mass ratio in (2.59).

These examples show that diagrams with electron loops, that contribute to the muon anomalous magnetic moment, are sensitive to loop momenta smaller than the mass of the muon. Such momenta configurations produce contributions that are often enhanced by either logarithms of the muon to electron mass ratio or large numerical factor (cf. π^2 in (2.66)). As we will show in the next section, such terms become of particular importance starting from the three-loop order in the perturbative expansion of the muon anomalous magnetic moment.

Consider now the vacuum polarization due to the τ lepton. In this case we have the opposite situation. Since $m_\tau \gg m$, we must study (2.50) in the limit $\sqrt{s} \sim m_\tau \gg m$. Using (2.53), we find in this limit,

$$a_1^\mu(s) \approx \left(\frac{\alpha}{\pi}\right)\frac{m^2}{3s}, \qquad s \gg m^2. \tag{2.67}$$

With this approximation, we derive

$$w_2^{\mathrm{vp},\tau} \approx \frac{m^2}{9}\int_{4m_\tau^2}^{\infty}\frac{\mathrm{d}s}{s^2}\sqrt{1 - \frac{4m_\tau^2}{s}}\left(1 + \frac{2m_\tau^2}{s}\right) = \frac{m^2}{45m_\tau^2}. \tag{2.68}$$

This result shows that the τ lepton contribution to a_μ is mass-suppressed in accord with the discussion in Sect. 2.2.

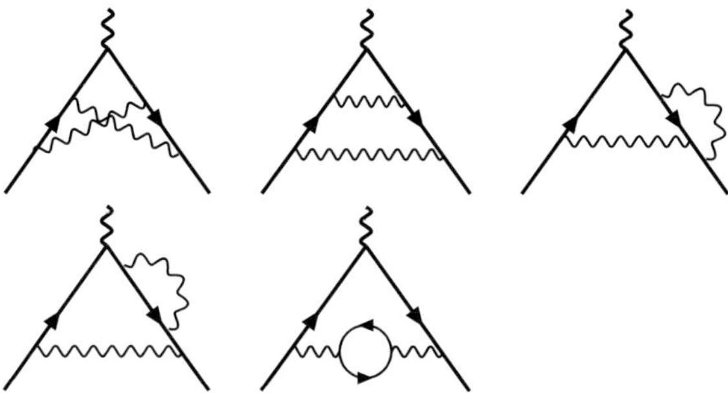

Fig. 2.4. Two-loop diagrams that contribute to w_2. Symmetric diagrams are not shown. The vacuum polarization diagram refers to the muon contribution only

The next step is the calculation of the term w_2 in (2.45). The corresponding Feynman diagrams are shown in Fig. 2.4. To compute w_2 we generalize the method discussed in connection with the computation of the Schwinger correction in the previous section. It is not difficult to see that all Feynman integrals needed to compute w_2 belong to one of the two classes

$$
S_2^{\mathrm{p}} = \int \frac{\pi^{-d} \mathrm{d}^d k_1 \mathrm{d}^d k_2}{k_1^{2a_1} k_2^{2a_2} (k_1 - k_2)^{2a_3} (k_1^2 + 2pk_1)^{a_4} (k_2^2 + 2pk_2)^{a_5}} \,,
$$

$$
S_2^{\mathrm{np}} = \int \frac{\pi^{-d} \mathrm{d}^d k_1 \mathrm{d}^d k_2}{k_1^{2a_1} k_2^{2a_2} (k_1^2 + 2pk_1)^{a_3} ((k_1 + k_2 + p)^2 - m^2)^{a_3} (k_2^2 + 2pk_2)^{a_5}} \,.
$$

(2.69)

Similar to the one-loop case, these integrals can be studied using identities obtained by the integration-by-parts method. There are six independent linear equations for each of $S_2^{\mathrm{p,np}}$ that follow from differentiating with respect to $k_{1,2}$ and contracting with $p, k_{1,2}$. These relations were studied in [19] where it was shown that any $S_2^{\mathrm{p,np}}$ integral can be expressed through three master integrals. We present these integrals as an expansion in $\epsilon = (4 - d)/2$ to the order that is needed in the calculation of the anomalous magnetic moment

$$
I_1 = S_2^{\mathrm{p}}(0,0,0,1,1) = S_1(0,1)\, S_1(0,1) \,,
$$

(2.70)

$$
I_2 = S_2^{\mathrm{p}}(1,0,1,1,0) = \Gamma^2(1+\epsilon) m^{2-4\epsilon} \left[-\frac{1}{2\epsilon^2} - \frac{5}{4\epsilon} - \frac{\pi^3}{3} - \frac{11}{8} \right.
$$
$$
\left. + \left(-\frac{5\pi^2}{6} - 4\zeta_3 + \frac{55}{16} \right) \epsilon \right] \,,
$$

(2.71)

$$
I_3 = S_2^{\mathrm{np}}(0,0,1,1,1) = \Gamma^2(1+\epsilon) m^{2-4\epsilon} \left[-\frac{3}{2\epsilon^2} - \frac{17}{4\epsilon} - \frac{59}{8} \right.
$$
$$
\left. + \left(-\frac{4\pi^2}{3} - \frac{65}{16} \right) \epsilon + \left(8\pi^2 \ln 2 - 28\zeta_3 + \frac{1117}{32} - \frac{26\pi^2}{3} \right) \epsilon^2 \right] \,,
$$

(2.72)

where $S_1(0,1)$ is given in (2.43).

In terms of those integrals, the contribution of the diagrams shown in Fig. 2.4 to w_2 reads

$$w_2^{\text{bare}} = \frac{1}{(4\pi)^{-2\epsilon}} \left[\frac{I_1}{m^4} \left(-\frac{3}{32\epsilon^2} + \frac{9}{32\epsilon} - \frac{45}{32} + \frac{41\epsilon}{96} + \frac{143\epsilon^2}{36} \right) \right.$$

$$+ \frac{I_2}{m^2} \left(\frac{1}{4\epsilon} + \frac{1}{8} - \frac{5\epsilon^2}{4} \right) \qquad (2.73)$$

$$\left. + \frac{I_3}{m^2} \left(-\frac{1}{16\epsilon^2} + \frac{5}{32\epsilon} - \frac{109}{96} + \frac{53\epsilon}{144} + \frac{305\epsilon^2}{216} \right) \right].$$

Using expressions for the master integrals Eq. (2.70, 2.71, 2.72), we find

$$w_2^{\text{bare}} = \frac{\Gamma^2(1+\epsilon)m^{-4\epsilon}}{(4\pi)^{-2\epsilon}} \left(\frac{23}{24\epsilon} + \frac{641}{144} + \frac{3}{4}\zeta_3 - \frac{\pi^2}{2}\ln 2 + \frac{\pi^2}{12} \right). \qquad (2.74)$$

The result is divergent. The finite result is obtained once the renormalization of the fine structure constant, the muon field and the muon mass are taken into account. We define

$$m_0 = Z_m m \,, \qquad \psi_0 = \sqrt{Z_2}\psi \,, \qquad \alpha_0 = \alpha Z_\alpha \,, \qquad (2.75)$$

where the subscript 0 denotes bare quantities. We emphasize that all the renormalization constants are computed in the on-shell scheme. Through $\mathcal{O}(\alpha)$ the renormalization constants read

$$Z_2 = Z_m = 1 + \left(\frac{\alpha}{\pi} \right) \frac{\Gamma(1+\epsilon)m^{-2\epsilon}}{(4\pi)^{-\epsilon}} \left(-\frac{3}{4\epsilon} - \frac{1}{1-2\epsilon} \right),$$

$$Z_\alpha = 1 + \left(\frac{\alpha}{\pi} \right) \frac{\Gamma(1+\epsilon)m^{-2e}}{3\epsilon(4\pi)^{-\epsilon}} \,. \qquad (2.76)$$

There are two types of counterterms that we have to consider. The first one is the insertion of the mass counterterm into the one-loop diagram Fig. 2.5. The result reads

$$\delta w_2^{\text{ct},1} = \frac{\Gamma(1+\epsilon)m^{-2\epsilon}}{(4\pi)^{-\epsilon}} \left(\frac{3}{4} + \epsilon \right) \frac{S_1(0,1)}{m^2} = \frac{\Gamma^2(1+\epsilon)m^{-4\epsilon}}{(4\pi)^{-2\epsilon}} \left(-\frac{3}{4\epsilon} - \frac{7}{4} \right). \qquad (2.77)$$

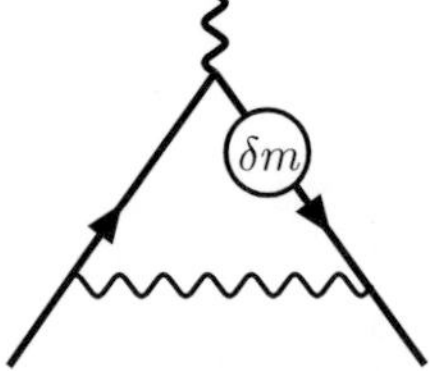

Fig. 2.5. The mass counter-term insertion into the one-loop diagram; the symmetric diagram is not shown

The second contribution comes from the multiplicative renormalization of the one-loop result; we write

$$a_\mu^{(1)} = \left(\frac{\alpha}{\pi}\right) Z_\alpha Z_2 \frac{\Gamma(1+\epsilon)m^{-2\epsilon}}{(4\pi)^{-\epsilon}}\left(\frac{1}{2}+2\epsilon\right), \tag{2.78}$$

from where we find

$$\delta w_2^{\mathrm{ct},2} = \frac{\Gamma^2(1+\epsilon)m^{-4\epsilon}}{(4\pi)^{-2\epsilon}}\left(-\frac{5}{24\epsilon}-\frac{4}{3}\right). \tag{2.79}$$

The final result for w_2 is obtained as the sum of (2.74, 2.77, 2.79). We derive

$$w_2 = \frac{197}{144}+\frac{3}{4}\zeta_3 - \frac{\pi^2}{2}\ln 2 + \frac{\pi^2}{12}, \tag{2.80}$$

where ζ_3 is given by the Riemann zeta-function $\zeta_p = \sum_{n=1}^{\infty}\frac{1}{n^p}$.

2.5 Three-loop QED Corrections to a_μ

At the third order in the perturbative expansion in QED, the technical difficulties become enormous. In addition, diagrams of the light-by-light scattering type, Fig.2.6, where, similar to hadronic vacuum polarization, all charged particles contribute, appear at this order. Traditionally, only light-by-light scattering diagrams with electron, muon and τ loops are included in the QED part of the muon magnetic anomaly. The light-by-light scattering diagrams mediated by the electron loops turn out to be particularly important; not only those diagrams are enhanced by $\ln m/m_e$, but also coefficients of the logarithms are large, $\sim \pi^2$. This interesting feature leads to a strong dominance of the light-by-light scattering contribution, mediated by the electron loop, in the three-loop QED contribution to a_μ, and indicates that similar diagrams are important in fourth and higher orders.

For our discussion, we split the three-loop QED correction to the muon anomalous magnetic moment into several components

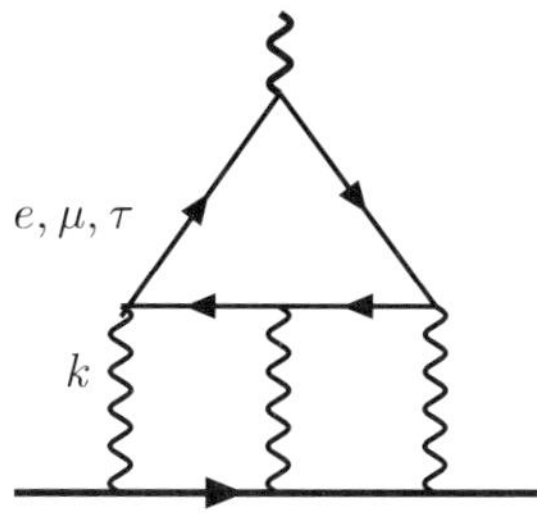

Fig. 2.6. The diagram with the light-by-light scattering loop

$$a_\mu^{(3)} = \left(\frac{\alpha}{\pi}\right)^3 \left[w_3^{\mathrm{lbl},e} + w_3^{\mathrm{lbl},\tau} + w_3^{\mathrm{vp}} + w_3\right], \tag{2.81}$$

where the abbreviations should be clear from the preceding discussion. The component w_3 accounts for all the diagrams that depend on the muon mass only.

We begin with the computation of w_3. Although numerically w_3 is smaller than the light-by-light and the vacuum polarization components, it required the largest effort to be derived analytically. The calculations started around 1970 [20] and were completed in 1995 by Laporta and Remiddi [21]. Because it took more than twenty years to finalize the project, different parts of the calculations were done using different techniques. From today's perspective, a convenient way to approach the problem is through the integration-by-parts identities for the three-loop massive on-shell integrals. To a certain extent, this approach was adopted by Laporta and Remiddi in the final stages of their work on w_3. A complete recursive solution of the integration-by-parts identities for the three-loop massive on-shell integrals can be found in [22, 23].

We briefly describe technical aspects of the calculation. There are forty distinct diagrams that contribute to w_3. To compute those, we first repeat all the steps discussed in connection with the one- and two-loop QED contributions to the muon anomalous magnetic moment. Upon removing the dependence on the momentum transfer q, w_3 is expressed through the three-loop on-shell massive integrals. Those integrals can be separated into eleven classes. For each of these classes, we write down a set of integration-by-parts identities and solve them iteratively. The solution shows that any on-shell three-loop massive integral is a linear combination of the seventeen master integrals. Analytic expressions for these integrals as well as some details about their evaluation can be found in [20, 21, 22]. The result reads

$$w_3 = \frac{83\pi^2}{72}\zeta_3 - \frac{215}{24}\zeta_5 + \frac{100}{3}\left[a_4 + \frac{\ln^4 2}{24} - \frac{\pi^2}{24}\ln^2 2\right] - \frac{239}{2160}\pi^4$$
$$+ \frac{139}{18}\zeta_3 - \frac{298}{9}\pi^2\ln 2 + \frac{17101}{810}\pi^2 + \frac{28259}{5184}, \tag{2.82}$$

where $a_4 = \sum_{n=1}^{\infty}\frac{1}{2^n n^4}$.

The second component of $a_\mu^{(3)}$ that we have to discuss is the vacuum polarization correction. Since it receives contributions from fermions of different masses, it is convenient to split it up further

$$w_3^{\mathrm{vp}} = w_3^{\mathrm{vp},e} + w_3^{\mathrm{vp},\tau} + w_3^{\mathrm{vp},e\tau}. \tag{2.83}$$

Here, $w_3^{\mathrm{vp},i}$ receives contributions from all diagrams that contain at least one vacuum polarization insertion of the fermion of type i. In addition $w_3^{\mathrm{vp},e\tau}$ refers to the diagram that contains both electron and τ vacuum polarization loops.

All contributions to w_3^{vp} are known analytically. The $w_3^{\mathrm{vp},e\tau}$ was obtained in [24] confirming earlier numerical result [25]. We do not present the full analytic result for $w_3^{\mathrm{vp},e\tau}$ since a simple estimate is sufficient, given the current level of precision for a_μ. The major three-loop effect related to the mixed vacuum polarization contribution is the combination of the two effects discussed in connection with the electron and τ vacuum polarization loops in the previous section; the electron loop changes the strength of the fine structure constant while the τ loop contributes m_μ^2/m_τ^2 power suppression term. If we write $a_\mu^{(2)} = (\alpha(m)/\pi)^2 m^2/(45m_\tau^2)$ for the two-loop τ vacuum polarization contribution and expand this expression in powers of α using (2.57), we obtain the $\ln(m/m_e)$ enhanced three-loop contribution to a_μ from the diagram with mixed electron-tau vacuum polarization loops

$$w_3^{\mathrm{vp},e\tau} \approx \frac{4m^2}{135m_\tau^2} \ln \frac{m}{m_e} \ . \tag{2.84}$$

The three-loop shift in a_μ induced by $w_3^{\mathrm{vp},e\tau}$ is $\sim 0.7 \times 10^{-11}$, well below the current experimental precision.

Similarly, the τ vacuum polarization contribution is very small. We give the corresponding result here for completeness

$$w_3^{\mathrm{vp},\tau} = \frac{m^2}{m_\tau^2} \left(-\frac{23}{135} \ln \frac{m_\tau}{m} - \frac{2\pi^2}{45} + \frac{10117}{24300} \right) . \tag{2.85}$$

More terms in the expansion of $w_3^{\mathrm{vp},\tau}$ in powers of m/m_τ can be found in [25].

The important vacuum polarization correction is again associated with the electron loops, $w_3^{\mathrm{vp},e}$; in this case the corrections enhanced by two powers of the large logarithm $\ln(m/m_e)$ are generated, as it is easy to understand from the running of the coupling constant, (2.57).

The result for $w_3^{\mathrm{vp},e}$ is known in an analytic form [25] but it is more convenient to present it as the series expansion in powers of m_e/m

$$
\begin{aligned}
w_3^{\mathrm{vp},e} =\ & \frac{2}{9} \ln^2 \frac{m}{m_e} + \left(\zeta_3 - \frac{2\pi^2}{3} \ln 2 + \frac{\pi^2}{9} + \frac{31}{27} \right) \ln \frac{m}{m_e} \\
& + \frac{11\pi^4}{216} - \frac{2\pi^2}{9} \ln^2 2 - \frac{8}{3} a_4 - \frac{\ln^4 2}{9} - 3\zeta_3 + \frac{5\pi^2}{3} \ln 2 - \frac{25\pi^2}{18} + \frac{1075}{216} \\
& + \frac{m_e}{m} \left(-\frac{13\pi^3}{18} - \frac{16\pi^2}{9} \ln 2 + \frac{3199\pi^2}{1080} \right) \\
& + \left(\frac{m_e}{m} \right)^2 \left(\frac{10}{3} \ln^2 \frac{m}{m_e} - \frac{11}{9} \ln \frac{m}{m_e} - \frac{14\pi^2}{3} \ln 2 - 2\zeta_3 + \frac{49\pi^2}{12} - \frac{131}{54} \right) \\
& + \left(\frac{m_e}{m} \right)^3 \left(\frac{4\pi^2}{3} \ln \frac{m}{m_e} + \frac{35\pi^3}{12} - \frac{16\pi^2}{3} \ln 2 - \frac{5771\pi^2}{1080} \right) + \mathcal{O} \left(\frac{m_e^4}{m^4} \right) .
\end{aligned} \tag{2.86}
$$

The $\mathcal{O}(\ln^2 m/m_e)$ term in (2.86) is easy to obtain from the renormalization group equation for the coupling constant, (2.56).

We now turn to the light-by-light scattering contributions due to electron and τ loops computed in [26]. The τ contribution is quite small. Taking the leading term from [26] we find

$$w_3^{\mathrm{lbl},\tau} = \left(\frac{3}{2}\zeta_3 - \frac{19}{16}\right)\frac{m^2}{m_\tau^2} + \mathcal{O}\left(\frac{m^4}{m_\tau^4}\right). \tag{2.87}$$

The corresponding shift in the anomalous magnetic moment is 2.7×10^{-11}, approximately.

On the contrary, the light-by-light scattering contribution mediated by the electron loop is very significant. The result derived in [26] reads

$$
\begin{aligned}
w_3^{\mathrm{lbl},e} = {} & \frac{2\pi^2}{3}\ln\frac{m}{m_e} + \frac{59\pi^4}{270} - 3\zeta_3 - \frac{10\pi^2}{3} + \frac{2}{3} \\
& + \frac{m_e}{m}\left(\frac{4\pi^2}{3}\ln\frac{m}{m_e} - \frac{196}{3}\pi^2\ln 2 + \frac{424}{9}\pi^2\right) \\
& + \left(\frac{m_e}{m}\right)^2\left(-\frac{2}{3}\ln^3\frac{m}{m_e} + \left(\frac{\pi^2}{9} - \frac{20}{3}\right)\ln^2\frac{m}{m_e}\right. \\
& \left. - \left(\frac{16}{135}\pi^4 + 4\zeta_3 - \frac{32\pi^2}{9} + \frac{61}{3}\right)\ln\frac{m}{m_e}\right. \\
& \left. + \frac{4\pi^2}{3}\zeta_3 - \frac{61\pi^4}{270} + 3\zeta_3 + \frac{25\pi^2}{18} - \frac{283}{12}\right) + \mathcal{O}\left(\frac{m_e^3}{m^3}\right).
\end{aligned}
\tag{2.88}
$$

An interesting feature of this result is the appearance of the logarithmically enhanced term with large coefficient $\sim \pi^2 \ln(m/m_e)$ Numerically, this term gives a large fraction of $w_3^{\mathrm{lbl},e}$ and originates [27] from the region of small loop momenta with the muon propagators on the mass shell. This mechanism of enhancing radiative corrections by factors of π was discussed in detail in the previous section.

Consider the integration over virtual momentum k in the light-by-light scattering diagram shown in Fig. 2.6 and assume that $m_e \ll k \ll m$. Since the virtual momentum is smaller than the mass of the muon, the muon serves as a source of time-independent electromagnetic field. This observation leads to the expectation that the logarithmically enhanced contribution might be enhanced by additional factors of π^2. Indeed, in the rest frame of the muon, its propagator reads

$$1/(k_0 + i0) \longrightarrow -i\pi\delta(k_0), \tag{2.89}$$

where the delta-function ensures that there is no energy transfer from the muon line to the electron light-by-light scattering loop. Simultaneously, we see that such a contribution has two additional factors of π, each coming from one of the static muon lines going on the mass shell. The calculation along these line was performed in [27]; although it is important for understanding the structure of $w_3^{\mathrm{lbl,e}}$, it is of little relevance numerically, since the full result is known. However, this observation is used to estimate the four- and five-loop QED corrections to a_μ, as we discuss in the next section.

2.6 Four- and Five-loop QED Corrections

The four-loop QED contribution to a_μ is the last known term in the perturbative expansion. We stress that, given the current experimental precision on the muon anomalous magnetic moment, the $\mathcal{O}(\alpha^4)$ correction is important for the comparison between theory and experiment.

The computation of the four-loop diagrams for a_μ represents the current frontier of perturbative calculations in Quantum Field Theory; it was being pursued by Kinoshita with collaborators over the last two decades. The most recent results are summarized in [28]. We comment on some aspects of the four-loop calculation below.

First, in contrast to calculations in lower orders, the four-loop computations are *fully* numerical; this concerns, for example, the renormalization of sub-divergences. Because of that, large cancelations between different pieces are involved and, as the authors of [28] point out, there appears the so-called digit deficiency problem that requires using **real*16** Fortran arithmetics to arrive at a stable result.

Second, so far there is no cross-check of the four-loop correction to a_μ by an independent group. Partial results [29] were used to elucidate some problems with earlier numerical work leading to the discovery of the digit deficiency problem. Currently, the calculation of $a_\mu^{(4)}$ is performed with two somewhat different methods and the results are consistent [28] .

Third, it is important to emphasize that, in spite of these problems, the four-loop QED contribution to the muon anomalous magnetic moment is known sufficiently well for phenomenological purposes. This happens because the four-loop correction to the muon magnetic anomaly is strongly dominated by the electron light-by-light scattering diagrams with the additional electron vacuum polarization insertion in one of the photon lines, Fig. 2.7. Those diagrams were computed independently by two groups [28, 30] with consistent results.

It is quite easy to understand why diagrams of the type shown in Fig. 2.7 dominate. Recall, that the most important contribution to the three-loop corrections discussed in the previous section comes from the electron light-by-light scattering diagram, Fig. 2.6; the contribution of this diagram is enhanced by the logarithm $\ln(m/m_e)$ with the large coefficient $\sim \pi^2$. This feature also makes the diagrams with the electron light-by-light scattering sub-loop dominant in the fourth order. Among the QED corrections to such diagrams, diagrams similar to Fig. 2.7 have an additional enhancement factor $\ln(m/m_e)$, so that the four-loop result can be estimated as

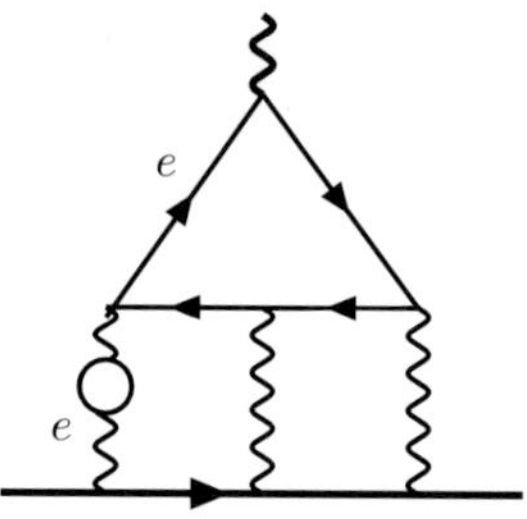

Fig. 2.7. The dominant four-loop diagram. Symmetric diagrams are not shown

$$a_\mu^{(4)} \sim \left(\frac{\alpha}{\pi}\right)^4 \frac{\pi^2}{2} \ln^2 \frac{m}{m_e} \, . \tag{2.90}$$

The four-loop contribution to the muon magnetic anomaly reads

$$a_4 = \left(\frac{\alpha}{\pi}\right)^4 \left(w_4^{\mathrm{lbl+vp},e} + w_4\right) , \tag{2.91}$$

where we neglect the contribution of the tau lepton at this order. The mass-independent term w_4 has been computed in [28]

$$w_4 = -1.7502(384) \, . \tag{2.92}$$

The term $w_4^{\mathrm{lbl+vp},\,e}$ that includes the electron loop in vacuum polarization and the light-by-light scattering diagrams reads

$$w_4^{\mathrm{lbl+vp},\,e} = 132.6823(72) \, . \tag{2.93}$$

This value should be compared with 117.4(5) [30], that is obtained upon evaluating the contribution to a_μ due to the diagrams shown in Fig. 2.7. It follows, that approximately ninety percent of $w_4^{\mathrm{lbl+vp},\,e}$ come from a few diagrams that are very well understood. It is precisely this feature of the four-loop result that makes it robust; this implies that any discrepancy between the experimental result for a_μ and the theoretical expectation at the level of 100×10^{-11} is very unlikely to be caused by deficiencies in QED computations.

The full calculation of the five-loop contribution to a_μ is not available. Nevertheless, it is reasonable to assume that electron light-by-light scattering diagrams dominate there and existing estimates of the five-loop contribution [27, 31, 32] are based on this assumption.

Following our explanation of how such diagrams contribute at the three- and four-loop level, it is easy to understand that there are two possibilities to obtain large contributions at the five-loop level. An additional electron loop insertion into one of the photons in the four-loop diagram Fig. 2.7 gives additional logarithms so that its contribution to w_5 can be estimated as $w_5 \sim \pi^2 \ln^3 m/m_e \approx 10^3$. Alternatively, the light-by-light scattering diagram with no electron vacuum polarization loops has four muon propagators that can go on-shell, hence $w_5 \sim \pi^4 \ln(m/m_e) \sim 500$. This discussion shows that a value of $w_5 \sim 10^3$ can be expected, which is confirmed by a more careful estimate [32]

$$w_5 = 930(170) \,. \tag{2.94}$$

Very recently an explicit computation of the leading contributions to w_5 was reported by Kinoshita [33]; he finds

$$w_5 = 677(40) \,. \tag{2.95}$$

We will use this value in our estimate of the full QED contribution to the muon magnetic anomaly discussed in the next section.

2.7 Complete QED Contribution to the Muon Magnetic Anomaly

In this section we combine the results discussed above to obtain complete QED contribution to the muon magnetic anomaly; for this we need the values of the fine structure constant α and the electron, muon and tau masses.

Given the experimental uncertainty on a_μ, the current precision on all those quantities is sufficient and their imprecise knowledge does not induce an appreciable uncertainty in the muon magnetic anomaly. We use values recommended by the National Institute of Standards [34],

$$\frac{1}{\alpha} = 137.03599911(46) \,, \quad \frac{m}{m_e} = 206.7682838(54) \,, \quad \frac{m_\tau}{m} = 16.8183(27) \,. \tag{2.96}$$

With this, we derive the following result for the QED contribution to the muon magnetic anomaly

$$a_\mu^{\mathrm{QED}} = 116\ 584\ 719(1) \times 10^{-11} \,, \tag{2.97}$$

where all QED effects through five-loops (2.95) have been included. Numerically, the five-loop corrections $\sim 6 \times 10^{-11}$ are not significant. Nevertheless, it became customary to include those terms into the estimate of a_μ^{QED}. The error in (2.97) is entirely due to the uncertainty in the five-loop contribution; the uncertainty caused by the imprecise knowledge of the input parameters is at the level of 10^{-12}.

References

1. V.B. Berestetskii, E.M. Lifshitz and L.P. Pitaevskii, *Quantum Electrodynamics*, Course in Theoretical Physics, vol. 4, Pergamon Press, 1982.
2. J. Schwinger, Phys. Rev. **76**, 790 (1949).

3. A history of the discovery of QED can be found in S. S. Schweber, *QED and the men who made it*, Princeton University Press, 1994.
4. T. Kinoshita, *Theory of the anomalous magnetic moment of the electron – numerical approach*, in [35].
5. V.B. Berestetskii, O.N. Krokhin and A.K. Klebnikov, Sov. Phys. JETP **3**, 761 (1956);
   W.S. Cowland, Nucl. Phys. B **8**, 397 (1958).
6. P. Kusch and H. M. Foley, Phys. Rev. **72**, 1256 (1947).
7. F.V. Tkachov, Phys. Lett. B **100**, 65 (1981);
   K.G. Chetyrkin and F.V. Tkachov, Nucl. Phys. B **192**, 345 (1981).
8. G. 't Hooft and M. Veltman, Nucl. Phys. B **44**, 189 (1972).
9. G. G. Bollini and J. J. Giambiagi, Phys. Lett. B **40**, 566 (1972);
   G. M. Cicuta and E. Montaldi, Nuovo Cimento Lett. **4**, 329 (1972);
   J. F. Ashmore, Nuovo Cimento Lett. **4**, 289 (1972).
10. W. J. Marciano and A. Sirlin, Nucl. Phys. B **88**, 86 (1975).
11. J. Collins, *Renormalization*, Cambridge University Press, 1984.
12. C. M. Sommerfeld, Phys. Rev. **107**, 328 (1957); Ann. Phys. (N.Y.), **5**, 26 (1958).
13. A. Petermann, Helv. Phys. Acta, **30**, 407 (1957); Nucl. Phys. **5**, 677 (1958);
    Phys. Rev. **105**, 1931 (1957).
14. H. Suura and E. Wichmann, Phys Rev. **105**, 1930 (1957).
15. H. H. Elend, Phys. Lett. **20**, 682 (1966); Erratum: *ibid*, **21**, 720 (1966).
16. M. Passera, J. Phys. G **31**, 75 (2005).
17. M. Gell-Mann, F. Low, Phys. Rev. **95**, 99 (1954).
18. For a review see, M. Neubert, Phys. Rept. **245**, 259 (1994); M. Shifman, Lectures at TASI 95, Boudler, Colorado [hep-ph/9510377].
19. D. J. Broadhurst, N. Gray and K. Schilcher, Zeit. f. Phys. C **52**, 111 (1991);
    N. Gray, D. J. Broadhurst, W. Grafe and K. Schilcher, Zeit. f. Phys. C **48**, 673 (1990).
20. See R. Roskies, M. Levine and E. Remiddi, *Analytic calculation of the sixth-order contribution to the electron's g factor*, in [35], for a review and reference to earlier work.
21. S. Laporta and E. Remiddi, Phys. Lett. B **379**, 283 (1996).
22. K. Melnikov and T. van Ritbergen, Nucl. Phys. B **591**, 515 (2000).
23. K. Melnikov and T. v. Ritbergen, Phys. Lett. B **482**, 99 (2000).
24. A. Czarnecki and M. Skrzypek, Phys. Lett. B **449**, 354 (1999).
25. S. Laporta, Nuovo Cimento A **106**, 675 (1993).
26. S. Laporta and E. Remiddi, Phys. Lett. B **301**, 440 (1993).
27. A. Yelkhovsky, *On the higher order QED contributions to the muon $g-2$ value*, preprint INP-88-50, (1988).
28. T. Kinoshita and M. Nio, Phys. Rev. D **70**, 113001 (2004); Phys. Rev. Lett. **90**, 021803 (2003).
29. P. Baikov and D. J. Broadhurst, in *New Computing Techniques in Physics Research IV*, proceedings of AIHEP, edited by B. Denby and D. Perret-Gallix, World Scientific, Singapore, p. 167 (1995) [hep-ph/9504398].
30. C. Chlouber and M. Samuel, Phys. Rev. D, **16**, 3596 (1977).
31. A. Milstein and A. Yelkhovsky, Phys. Lett. B **233**, 11 (1989).
32. S. Karshenboim, Phys. Atom. Nucl., **56**, 857 (1993).
33. T. Kinoshita, Nucl. Phys. Proc. Suppl. **144**, 206 (2005).
34. See http://physics.nist.gov/constants.
35. T. Kinoshita (ed), *Quantum Electrodynamics*, World Scientific, 1990.

3 Hadronic Vacuum Polarization

3.1 Hadronic Vacuum Polarization: The Basics

The hadronic vacuum polarization, Fig. 3.1, is the largest hadronic effect on the muon anomalous magnetic moment; it has to be known with high precision to match the existing experimental effort. Given the low value of the muon mass, this task seems formidable because at low energies hadronic interactions are strong. We are rescued by a dispersion representation of the photon propagator that relates hadronic vacuum polarization contribution to a_μ and the experimentally measured e^+e^- annihilation cross section into hadrons. In principle, this allows to account for the effects of strong interactions *exactly*. Such an approach puts, however, a significant burden on the experiment, requiring high precision measurement of the e^+e^- annihilation cross section.

We begin by giving a crude estimate of the hadronic vacuum polarization contribution to a_μ. Following the discussion in Sect. 2.4, it is easy to see that a hadronic state with the invariant mass M_{hadr} changes a_μ by

$$a_\mu^{\mathrm{hvp}} \sim \left(\frac{\alpha}{\pi}\right)^2 \frac{m^2}{M_{\mathrm{hard}}^2} \, . \tag{3.1}$$

Taking $M_{\mathrm{hard}} \sim 1$ GeV as a typical scale for hadron masses, we arrive at the estimate

$$a_\mu^{\mathrm{hvp}} \sim 6000 \times 10^{-11} . \tag{3.2}$$

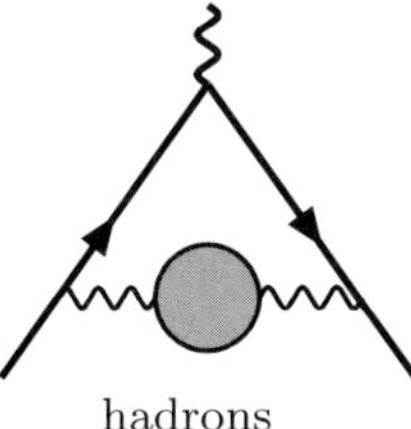

Fig. 3.1. The hadronic vacuum polarization contribution to the muon anomalous magnetic moment

K. Melnikov and A. Vainshtein: *Theory of the Muon Anomalous Magnetic Moment*
STMP **216**, 33–87 (2006)
© Springer-Verlag Berlin Heidelberg 2006

Although this estimate is in the correct range, one may wonder if pions, whose masses are close to the mass of the muon, change this estimate significantly. We describe a refined estimate in Sect. 3.3 where we demonstrate that the contribution to a_μ^{hvp} of the two-pion state with an invariant mass close to its minimal value $2m_\pi$ is fairly small indeed.

To derive a useful representation for the hadronic vacuum polarization contribution to the muon anomalous magnetic moment, we consider Feynman diagram Fig. 3.1. This diagram can be computed by following the steps described in Sect. 2.4 in connection with the computation of the lepton vacuum polarization contribution to a_μ. An obvious modification of (2.50) leads to the result

$$a_\mu^{\mathrm{hvp}} = \frac{\alpha}{3\pi} \int\limits_{s_0}^{\infty} \frac{\mathrm{d}s}{s} \, R^{\mathrm{hadr}}(s) \, a_\mu^{(1)}(s) \, , \tag{3.3}$$

where $a_\mu^{(1)}(s)$ can be found in (2.51) and

$$R^{\mathrm{hadr}}(s) = \frac{\sigma_{\mathrm{hadr}}(s)}{\sigma_{\mathrm{point}}(s)} \, , \tag{3.4}$$

with σ_{point} given in (2.49) and σ_{hadr} being the $e^+ e^-$ hadronic annihilation cross section.

In (3.3), s_0 refers to the lowest invariant mass squared of the hadronic system that can be produced in the $e^+ e^-$ annihilation. We use $s_0 = 4m_\pi^2$, the threshold for the production of the two pions; however, strictly speaking, $s_0 = 0$, since such "hadronic" states as 3γ and $\pi^0 \gamma$ are produced in the $e^+ e^-$ annihilation as well. In practice, the contribution of the energy region $0 \leq s \leq 4m_\pi^2$ to a_μ^{hvp} is less than 1×10^{-11} [1] and, therefore, can be safely neglected.

It is easy to see that the integral in (3.3) is dominated by hadronic states with relatively low invariant masses. Indeed, for large s, $R^{\mathrm{hadr}}(s)$ becomes s-independent, whereas $a_\mu^{(1)}(s) \sim m^2/s$. Hence, for $s \gg m^2$, the integrand in (3.3) decreases as $1/s^2$ and the integral converges rapidly.

Equations (3.3, 3.4) as well as (2.51) are sufficient to compute the hadronic vacuum polarization contribution to the muon anomalous magnetic moment; given the experimental result for $\sigma_{\mathrm{hadr}}(s)$, it is straightforward to evaluate the integral in (3.3) and obtain a_μ^{hvp}. In practice, however, this task is complicated by the need to combine data from many experiments and analyze systematic uncertainties associated with each data set.

Before discussing these issues in detail, in Sect. 3.3 we describe a way to estimate a_μ^{hvp}. Such an estimate is instructive for two reasons. It allows us to develop an intuitive understanding of the physics behind the hadronic vacuum polarization contribution to the muon anomalous magnetic moment and, in addition, to introduce theoretical arguments that are used in Chaps. 5, 6 where other hadronic contributions are discussed. The theoretical tools that we extensively use in the discussion of hadronic effects in the muon magnetic anomaly, are the operator product expansion and the limit of QCD where the number of colors N_c is considered large; we describe them in the next section.

3.2 Theoretical Methods

3.2.1 Large-N_c Approximation in QCD

In QCD the number of colors N_c equals three. However, it was pointed out by 't Hooft [2] that generalizing SU(3) gauge group of QCD to SU(N_c) and considering N_c as a free parameter, provides a useful tool for the theoretical analysis of non-perturbative aspects of QCD. In particular, 't Hooft showed that in the limit $N_c \to \infty$ with $N_c \alpha_s$ fixed, the two-dimensional QCD can be solved exactly. In spite of significant effort, the four-dimensional generalization of this solution has not yet been found. Nevertheless, important general features of large-N_c QCD in four dimensions are known.

The key to understanding large-N_c QCD are the counting rules that establish dependence of Green's functions on N_c. A detailed discussion of these rules can be found in [2, 3]; here we briefly sketch it. Starting from the original QCD Lagrangian

$$\mathcal{L} = -\frac{1}{4g_0^2}\, G_{\mu\nu}^a G^{a,\mu\nu} + \bar{\psi}\big(i\hat{D} - M\big)\psi\,, \qquad (3.5)$$

where ψ denotes the quark field with its flavor, color and Dirac indices and M is the diagonal mass matrix, $M = \mathrm{Diag}(m_u, m_d, \ldots)$, and redefining the coupling constant $g_0 \to g/\sqrt{N_c}$ and the quark field $\psi = \sqrt{N_c}\tilde{\psi}$, the Lagrangian is cast into a form suitable for the large-N_c analysis

$$\mathcal{L} = N_c \left(-\frac{1}{4g^2}\, G_{\mu\nu}^a G^{a,\mu\nu} + \bar{\tilde{\psi}}\big(i\hat{D} - M\big)\tilde{\psi} \right). \qquad (3.6)$$

This form of the Lagrangian implies that, in a given Feynman diagram, each propagator contributes a factor $1/N_c$ and each vertex contributes a factor N_c, to an overall N_c-scaling.

This is not the whole story, however, since internal lines in Feynman diagrams are charged under SU(N_c) and, therefore, sums over internal color indices generate powers of N_c. A useful way to determine the leading power of N_c that originates from such sums is the double-line notation for the gluon propagator, suggested by 't Hooft [2]. In this picture, a gluon line in a Feynman diagram is represented by a pair of lines, one that corresponds to a color index and the other one that corresponds to an anti-color index; once the gluon line splits into a quark and an anti-quark lines, the color index of a gluon is absorbed by a quark, and an anti-color index, by an anti-quark. Such representation turns an ordinary Feynman diagram with quark and gluon lines into a collection of polygons glued together to form surfaces; each closed color loop forms the edge of the polygon and is the face of the surface. The total number of edges is equal to the number of propagators in a given Feynman diagram. Each color loop provides a single power of N_c to an overall N_c-scaling. Hence, the leading N_c-dependence of a Feynman diagram with V vertices, E edges and F faces is

$$N_c^{V-E+F} = N_c^{\chi} \,, \tag{3.7}$$

The parameter χ is a topological invariant of a two-dimensional surface that is obtained from a Feynman diagram; it is called the Euler character. Instead of vertices, edges and faces, the Euler character can be expressed through the number of *holes (boundaries)* b and *handles* h, of a two-dimensional surface that corresponds to a given Feynman diagram

$$\chi = 2 - 2h - b \,. \tag{3.8}$$

The maximal value of χ is two; it is obtained for a two-dimensional surface without holes and handles – the sphere.

The fact that the leading power of N_c is related to a topological property of a Feynman diagram, permits an easy proof of the following powerful result. The leading contribution to connected vacuum-to-vacuum transition amplitude scales like N_c^2 and corresponds to *planar* diagrams made up of gluons only. Planar diagrams with an additional quark loop have a single boundary $b = 1$; hence $\chi = 1$ and the scaling is N_c. The quark loop has to form a boundary of the graph. Note, that the above analysis and the power-counting applies to the re-scaled coupling constant g and the quark field $\tilde{\psi}$.

We will be particulary interested in what large-N_c QCD has to say about properties of hadrons; we assume that confinement persists in the large-N_c limit and, hence, analysis in terms of hadrons is appropriate. To analyze meson properties, we require Green's functions composed of quark bilinears with the quantum numbers that correspond to mesons of interest. Considering the vacuum-to-vacuum transition amplitude in the presence of external currents and using results for the vacuum-to-vacuum transition amplitude described above, it is easy to derive the large-N_c counting rule for the Green's functions composed of quark bilinears. Denoting such bilinears by $\tilde{J} \sim \bar{\tilde{\psi}} \Gamma_J \tilde{\psi}$, where the matrix Γ_J is a generic notation for the Lorentz and flavor structure of the current, we derive

$$\langle \tilde{J}_1 \tilde{J}_2 .... \tilde{J}_n \rangle \propto N_c^{1-n} \,, \tag{3.9}$$

where $\langle ... \rangle$ denotes the vacuum average of the time-ordered product.

We can understand some interesting features of the large-N_c QCD using (3.9). First, it implies an N_c-scaling rule for meson-meson coupling constants. Consider, for example, the $\rho\pi\pi$ vertex. The coupling constant can be obtained from the Green's function of the interpolating currents for the ρ meson and the two pions. In terms of non-rescaled quark fields ψ, these interpolating currents have the form $\bar{\psi}\Gamma_J\psi/\sqrt{N_c}$ since mesons are color singlets and their wave functions are normalized in an N_c-independent way. Hence, the coupling constant is deduced from the following set of equations

$$g_{\rho\pi\pi} \; \propto \; \frac{1}{N_c^{3/2}}\langle \bar{\psi}\Gamma_\rho\psi \, \bar{\psi}\Gamma_\pi\psi \, \bar{\psi}\Gamma_\pi\psi\rangle = N_c^{3/2}\langle \bar{\tilde{\psi}}\Gamma_\rho\tilde{\psi}\,\bar{\tilde{\psi}}\Gamma_\pi\tilde{\psi}\,\bar{\tilde{\psi}}\Gamma_\pi\tilde{\psi}\rangle$$

$$\propto N_c^{3/2}N_c^{1-3} \propto \frac{1}{\sqrt{N_c}}\,. \tag{3.10}$$

This result is, of course, more general than just the $\rho\pi\pi$ coupling as can be seen from the derivation. Hence, we conclude that interactions between mesons in the large-N_c limit vanish, which implies that all mesons in that limit are *stable*.

In a similar fashion, we can estimate the N_c-scaling of Green's functions that involve external quark currents. Consider, for example, the axial current $\bar{\psi}T^-\gamma_\mu\gamma_5\psi = \bar{d}\gamma_\mu\gamma_5 u$, whose matrix element between the vacuum and the pion gives the pion decay constant F_π. Using (3.9), we derive the following scaling

$$F_\pi \propto \langle \bar{\psi}T^-\gamma_\mu\gamma_5\psi \, \frac{1}{\sqrt{N_c}}\, \bar{\psi}\Gamma_\pi\psi\rangle = N_c^{3/2}\langle \bar{\tilde{\psi}}T^-\gamma_\mu\gamma_5\tilde{\psi}\,\bar{\tilde{\psi}}\Gamma_\pi\tilde{\psi}\rangle \propto \sqrt{N_c}\,. \tag{3.11}$$

Clearly, nothing in this result is particular to the pion decay constant; any matrix element of that type has similar scaling. An important case for our purposes is the contribution of the ρ-meson to the e^+e^- annihilation cross section. Such contribution is proportional to the square of the matrix element of the electromagnetic current J_μ^{em} between the vacuum and the ρ-meson, $\langle 0|J_\mu^{\mathrm{em}}(0)|\rho\rangle = m_\rho g_\rho \epsilon_\mu^\rho$. Generalizing (3.11), we find that the contribution of the ρ-meson to the hadronic e^+e^- annihilation cross section is proportional to N_c. In addition to the ρ-meson contribution to the e^+e^- annihilation cross section, we also require the $\pi^+\pi^-$ contribution in what follows. It is easy to estimate it using the example above; as compared to (3.11), the presence of an additional pion leads to an extra factor of $1/\sqrt{N_c}$. Hence, we find that the $e^+e^- \to \pi^+\pi^-$ annihilation cross section is N_c-independent.

The results discussed above can be used to constrain the functional form of Green's functions in the large-N_c QCD. As an example, consider the correlation function of two electromagnetic hadronic currents $\langle J_\mu^{\mathrm{em}} J_\nu^{\mathrm{em}}\rangle$; we are interested in this quantity since its imaginary part gives the e^+e^- annihilation cross section into hadrons. The large-N_c counting rules discussed above imply that $\langle J_\mu^{\mathrm{em}} J_\nu^{\mathrm{em}}\rangle \sim N_c$. We may try to understand this result by considering contributions of mesons to this Green's function. Then, from the discussion above it follows that a *single* meson contribution to $\langle J_\mu^{\mathrm{em}} J_\nu^{\mathrm{em}}\rangle$ scales like N_c, the contribution of an intermediate two-meson state scales like $N_c^0 \sim 1$ and contributions of intermediate states with even higher meson multiplicities are even stronger suppressed. Hence, we conclude that in the large-N_c limit only one-meson intermediate states contribute to the correlation function of two electromagnetic currents. We write

$$i \int \mathrm{d}^4 x e^{iqx} \langle T\{J_\mu^{\mathrm{em}}(x) J_\nu^{\mathrm{em}}(0)\}\rangle = (q_\mu q_\nu - g_{\mu\nu} q^2) \sum_i^\infty \frac{g_i^2}{m_i^2 - q^2} \,, \qquad (3.12)$$

where $g_i^2 \propto N_c$ refers to the coupling of i-th vector meson to the electromagnetic current. It is easy to see that the resulting linear N_c dependence holds for any number of external currents.

In contrast to large-N_c QCD spectrum, real-world mesons are unstable. However, this difference becomes largely irrelevant once integrals of Green's functions are considered. A good example is a determination of meson properties from QCD sum rules when mesons are often approximated by delta-functions $\delta(s - M^2)$ in phenomenological spectral densities.

We note that Green's functions constructed as infinite sums over meson propagators have to match perturbative QCD description at short distances. Once this is achieved, we obtain the Green's function that satisfies basic QCD constrains and is defined for all values of q^2 because of the large-N_c Anzats. We use this approach extensively to study strong interaction effects in the muon magnetic anomaly.

3.2.2 The Operator Product Expansion

The operator product expansion (OPE) states that in Quantum Field Theory a time-ordered product of two local operators $A(y)$ and $B(y + x)$ can be expanded into a sum of local operators $\mathcal{O}_i(y)$ in the limit $x \to 0$,

$$T\{A(x)B(y + x)\} = \sum_i c^i(x)\mathcal{O}_i(y) \,, \qquad (3.13)$$

where summation over i may include Lorentz indices. The OPE was suggested by Wilson in 1969 [4]. Since then it found applications in a variety of topics in high energy physics, from factorization in deep-inelastic scattering and weak nonleptonic decays of K and π mesons and hyperons, to QCD sum rules and the theory of B meson decays. The OPE helps to separate contributions of large and small distances or, alternatively, small and large momenta, to a given physical process and to deal with the two components almost separately. We introduce the operator product expansion through an example, relevant for our discussion of the hadronic vacuum polarization. Our presentation follows original publications [5] where the hadronic vacuum polarization was studied in the context of QCD sum rules.

Consider the electromagnetic hadronic current

$$J_\mu^{\mathrm{em}}(x) = \sum_{q=u,d,s} Q_q \bar{q}(x) \gamma_\mu q(x) \,, \qquad (3.14)$$

where Q_q stands for the electric charge of the quark q in units of the positron charge $|e|$. We are interested in the Fourier transform of the product of two such currents

$$\hat{\Pi}_{\mu\nu} = i \int d^4x e^{iqx} T \left\{ J_\mu^{\mathrm{em}}(x) J_\nu^{\mathrm{em}}(0) \right\} , \qquad (3.15)$$

in the limit when momentum q is large and space-like, $q^2 = -Q^2$, $Q^2 \gg \Lambda_{\mathrm{QCD}}^2$. In this limit $\hat{\Pi}_{\mu\nu}(q^2)$ in (3.15) receives appreciable contribution from the small integration region, $x \sim 1/Q \to 0$; the contribution of the integration region where $x \gg 1/Q$ cancels out due to a rapidly oscillating exponential factor. Then, the general OPE expansion (3.13) leads to

$$\hat{\Pi}_{\mu\nu} = \sum_i c^i_{\mu\nu\alpha_1\ldots\alpha_i}(q)\mathcal{O}_i^{\alpha_1\ldots\alpha_i}(0) . \qquad (3.16)$$

In this equation the OPE coefficients $c^i_{\mu\nu\alpha_1\ldots\alpha_i}(q)$ are some c-valued tensors that depend on q.

Indices α_k of the operator $\mathcal{O}_i^{\alpha_1\ldots\alpha_i}$ are associated with its Lorentz spin; Lorentz spins of relevant operators depend on a particular process under consideration. For example, when the OPE is applied to deep inelastic processes, we average $\mathcal{O}_i^{\alpha_1\ldots\alpha_i}$ over hadronic target so that the momentum and the spin of the target define the tensor structure. In case of the vacuum polarization, we average (3.16) over the vacuum state; as a consequence, only operators that are scalars under Lorentz transformations contribute. These operators have no uncontracted indices and the tensor structure of their coefficients,

$$c^i_{\mu\nu}(q) = (q_\mu q_\nu - g_{\mu\nu}q^2)c^i(Q^2) , \qquad (3.17)$$

is fixed by the Lorentz invariance and the conservation of the electromagnetic current. Hence, the spin zero part of the OPE can be written as

$$\hat{\Pi}_{\mu\nu} = (q_\mu q_\nu - g_{\mu\nu}q^2)\hat{\Pi}(Q^2) , \qquad \hat{\Pi}(Q^2) = \sum_i c^i(Q^2)\mathcal{O}_i(0) . \qquad (3.18)$$

Within the OPE, the coefficient functions $c^i(Q^2)$ are determined by short distances and, consequently, large virtual momenta, while physics at large distances enters matrix elements of operators $\mathcal{O}_i$. To make the separation between the operators and the coefficient functions unambiguous, we require an introduction of the normalization point μ. Then, the coefficient functions c^i account for the contribution of virtual momenta larger than μ while momenta that are smaller than μ, contribute to the matrix elements of operators. Because physical results have to be μ-independent, the μ-dependence of the OPE coefficient functions is canceled by the μ-dependence of the matrix elements.

The features of the operator product expansion that are discussed above, are quite general; they are valid in any theory. Specific to QCD is the fact that the coupling constant decreases with momentum growth [6, 7], $\alpha_s(Q) \sim$

$1/\ln(Q/\Lambda_{\mathrm{QCD}}) \ll 1$ when $Q \gg \Lambda_{\mathrm{QCD}}$. For large values of Q, this feature allows the computation of the dominant part of the coefficient functions in perturbation theory.

Because $\hat{\Pi}(Q^2)$ is dimensionless, the mass dimension of the operator $\mathcal{O}_i$ determines the power behavior of the corresponding coefficient function. Denoting the mass dimension of the operator $\mathcal{O}_i$ by d_i, we derive

$$c_i(Q^2) = \frac{\tilde{c}}{Q^{d_i}} \ . \tag{3.19}$$

We can construct local operators by taking products of quantum fields and covariant derivatives; since each of these ingredients has positive mass dimension, more complicated operators have larger mass dimensions and, hence, give smaller contributions to $\hat{\Pi}(Q^2)$. Therefore, for a sufficiently large value of Q^2, there is just a handful of operators with *smallest* mass dimension that should be accounted for.

In addition to canonical mass dimensions, operators are also characterized by anomalous dimensions γ_i, which determine the dependence of operators on the normalization point μ. This dependence is governed by the renormalization group evolution equations [8] and is of the form $(\ln\mu)^{\gamma_i}$. Since μ-independent physical quantities are given by products of coefficient functions and matrix elements of operators, the dependence of the coefficient functions on μ can be deduced,

$$\tilde{c}_i \propto \left(\ln\frac{Q}{\mu}\right)^{-\gamma_i} . \tag{3.20}$$

The operator product expansion is important because it allows us to describe physical processes beyond the perturbation theory [5]. Indeed, in our example, the vacuum average of the product of two electromagnetic currents gives the hadronic vacuum polarization

$$\Pi(Q^2) = \langle 0|\hat{\Pi}(Q^2)|0\rangle = \sum \frac{\tilde{c}^i}{Q^{d_i}} \langle\mathcal{O}_i\rangle \ . \tag{3.21}$$

The vacuum averages $\langle\mathcal{O}_i\rangle$, that are also referred to as *condensates*, scale as $\Lambda_{\mathrm{QCD}}^{d_i}$ where d_i is the canonical mass dimension of the operator $\mathcal{O}_i$. Hence, as we explained earlier, taking $Q \gg \Lambda_{\mathrm{QCD}}$ allows us to truncate the sum in (3.21) and consider a limited number of local operators.

A word of caution is appropriate here. In the simplified picture presented above, the factorization of short and large distance physics into the OPE coefficient functions and the operators is equivalent to the factorization of perturbative and non-perturbative phenomena; the coefficient functions are given by the perturbative expansion in $\alpha_s(Q)$ while the matrix elements, proportional to powers of $\Lambda_{\mathrm{QCD}} = Q\exp[-\mathrm{Const}/\alpha_s(Q)]$, are non-perturbative. In reality, things are somewhat different. First, because of the μ-dependence, the matrix elements contain perturbative contributions. Second, nonperturbative physics exists even at short distances; small-size instantons serve as an example. Such contributions show up as power corrections to the OPE coefficient functions, of the form $\Lambda_{\mathrm{QCD}}^2/Q^2$, providing a new source of non-perturbative phenomena. In practical applications it is often *assumed* that the non-perturbative part of the OPE coefficient functions is numerically small and can be neglected; however, this may not always be a safe assumption.

We now list local operators that contribute to the right hand side of (3.21). As we already know, we have to start with those operators that have the lowest possible mass dimension. Because we take the matrix elements between the two vacuum states, the operators should be Lorentz scalars; in addition, they should be gauge invariant.

The simplest operator is the identity operator $\mathcal{O}_0 = \hat{\mathbf{I}}$, with the mass dimension zero. To construct operators of higher mass dimension, we should consider quark and gluon fields as well as covariant derivatives. The quark field has the mass dimension $3/2$, the covariant derivative – the mass dimension one and the gluon field-strength tensor $G_{\mu\nu}$ – the mass dimension two. Then, it is easy to see that the identity operator is followed by the dimension-three operators $\bar{q}q$ where $q = u, d, s$ and contraction of color indices is implied. In the chiral limit, $m_q = 0$, the OPE coefficient functions of these operators vanish because of the chiral properties of $\bar{q}q$. Indeed, these operators involve a helicity flip, $\bar{q}q = \bar{q}_R q_L + \bar{q}_L q_R$, while the electromagnetic currents preserve helicity, $\bar{q}\gamma_\mu q = \bar{q}_L \gamma_\mu q_L + \bar{q}_R \gamma_\mu q_R$. When small quark masses are introduced, the operators $\bar{q}q$ appear in the OPE in the combination

$$\mathcal{O}_4^{\mathrm{q}} = \sum_{q=u,d,s} Q_q^2 m_q \bar{q}q \ . \tag{3.22}$$

The mass dimension of the operator $\mathcal{O}_4^{\mathrm{q}}$ is four; this implies that the corresponding OPE coefficient function is proportional to $1/Q^4$. Note that the anomalous dimension of the operator $\mathcal{O}_4^{\mathrm{q}}$, (3.22), is zero since the μ-dependence of $m_q(\mu)$ is canceled by the μ-dependence of the operator $\bar{q}q$. There is no other dimension-four operator composed of the quark fields; the operator $\bar{q}\gamma^\mu D_\mu q$ reduces to $m_q \bar{q}q$ by equations of motion. However, an additional $d = 4$ operator can be composed using the gluon field-strength tensors

$$\mathcal{O}_4^{\mathrm{gl}} = \frac{\alpha_s}{\pi} \, G_{\mu\nu}^a G^{a,\mu\nu}. \tag{3.23}$$

The coupling constant α_s is included into the definition of $\mathcal{O}_4^{\mathrm{gl}}$ for convenience since, in the leading logarithmic approximation, the operator $\mathcal{O}_4^{\mathrm{gl}}$ is independent of the normalization scale μ. Hence, the operator product expansion for $\Pi(Q^2)$ reads

$$\Pi(Q^2) = c_0 + \frac{\tilde{c}_q}{Q^4} \left\langle \mathcal{O}_4^{\mathrm{q}} \right\rangle + \frac{\tilde{c}_{\mathrm{gl}}}{Q^4} \left\langle \mathcal{O}_4^{\mathrm{gl}} \right\rangle + \mathcal{O}\left(Q^{-6}\right). \tag{3.24}$$

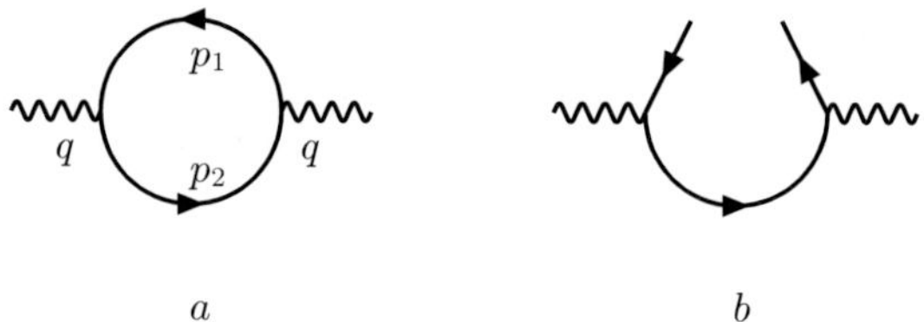

Fig. 3.2. Sample diagrams for computing the operator product expansion of two electromagnetic currents

An interesting feature of the OPE of the correlator of two electromagnetic currents, (3.24), is the absence of the $1/Q^2$ correction which follows from the fact that no dimension two Lorentz invariant local operator exists in QCD.[1] We use this feature in Sect. 3.3 to construct the theoretical model for the hadronic vacuum polarization contribution to the muon anomalous magnetic moment.

We now turn to the discussion of how the coefficient functions in the OPE can be computed. The key observation is that (3.18) is the operator equality, which implies that we can take matrix elements of both sides of that equation with respect to any set of states. A convenient set of states is provided by the perturbation theory; it contains perturbative vacuum, quarks and gluons. Upon considering the matrix elements of (3.18) with respect to those states, we find the coefficient functions of the corresponding operators.

As an example, consider the calculation of the coefficient function of the identity operator. To compute its coefficient function c_0, we have to average the time-ordered product of the two electromagnetic currents over the *perturbative* vacuum state $|0)$, different from the true vacuum $|0\rangle$. In perturbation theory, this correlator appears at the one-loop order and is given by a familiar diagram, Fig. 3.2a, while from the OPE perspective $(0|\hat{\Pi}_{\mu\nu}|0) = c_0$ because matrix elements of operators of higher mass dimension with respect to perturbative vacuum vanish thanks to our normalization conditions. Hence, a straightforward computation of the one-loop diagram Fig. 3.2a leads to the coefficient function of the unity operator c_0

$$c_0 = \frac{N_c}{12\pi^2} \sum_q Q_q^2 \ln\left(\frac{M_{\mathrm{UV}}^2}{Q^2}\right) , \qquad (3.25)$$

where $N_c = 3$ is the number of colors and M_{UV} is the ultra-violet cut-off. Because, in what follows, we consider $d\Pi/dQ^2$, the divergent constant in c_0 will be of no relevance for us.

Another example is the calculation of the coefficient function of the operator $\mathcal{O}_q$. To this end, we consider the average $(p|\hat{\Pi}_{\mu\nu}(q)|p)$ of the time-ordered product of two electromagnetic currents in (3.15) over the perturbative

---

[1] The absence of the $1/Q^2$ correction is related to the assumption that short-distance contributions to c_0 do not generate non-perturbative $\Lambda_{\mathrm{QCD}}^2/Q^2$ terms.

one-quark state $|p\rangle$ described by the Dirac spinor u_p. We assume that the quark momentum p and the quark mass m_q are much smaller than the momentum q. The average is given by the forward Compton scattering amplitude, Fig. 3.2b, and reads

$$\langle p|\hat{\Pi}_{\mu\nu}(q)|p\rangle = -Q_q^2 \bar{u}_p \left(\gamma_\mu \frac{1}{\hat{q} + \hat{p} - m_q} \gamma_\nu - \gamma_\nu \frac{1}{\hat{q} - \hat{p} + m_q} \gamma_\mu \right) u_p \ . \quad (3.26)$$

On the other hand, this matrix element can be computed from the operator product expansion

$$\langle p|\hat{\Pi}_{\mu\nu}(q)|p\rangle = (q_\mu q_\nu - g_{\mu\nu}q^2) \, \frac{\tilde{c}_q}{q^4} \, \langle p|Q_q^2 m_q \bar{q}q|p\rangle$$

$$= (q_\mu q_\nu - g_{\mu\nu}q^2) \, \frac{\tilde{c}_q}{q^4} \, Q_q^2 m_q \bar{u}_p u_p \ . \quad (3.27)$$

Equations (3.26), (3.27) should coincide in the limit $q \gg p, m_q$ for Lorentz scalar operators. The comparison between the two equations can be easily accomplished if we expand (3.26) in powers of $p/q, m_q/q$, contract both equations with $g^{\mu\nu}$ and average over directions of the vector q. From (3.27) we obtain

$$\langle p|\hat{\Pi}_\mu^\mu(q)|p\rangle = -3\tilde{c}_q \, \frac{m_q Q_q^2}{q^2} \, \bar{u}_p u_p \ , \quad (3.28)$$

while simple algebraic manipulations with (3.26) lead to

$$\langle p|\hat{\Pi}_\mu^\mu(q)|p\rangle = -6 \, \frac{m_q Q_q^2}{q^2} \, \bar{u}_p u_p \ . \quad (3.29)$$

Comparing (3.28, 3.29), we derive the coefficient function $\tilde{c}_q = 2$.

The coefficient functions can also be computed in a related, but slightly different way that elucidates important physics of the operator product expansion. It is based on the analysis of momentum flow in Feynman diagrams. Consider the one-loop Feynman diagram that describes the correlator of two electromagnetic currents Fig. 3.2a. The large momentum q can flow through this diagram in three different ways; $i)$ $p_1 \sim p_2 \sim Q$; $ii)$ $p_1 \ll p_2 \sim Q$ and $iii)$ $p_2 \ll p_1 \sim Q$. In the first case, all the quark lines are far off-shell and perturbative computations are applicable. This momentum region gives the coefficient function of the unity operator, (3.25). To investigate the contribution of the two regions where one quark line is soft, we proceed as follows. Writing the expression for the one-loop contribution to the current correlator in momentum space and contracting it with $g^{\mu\nu}$ we obtain

$$3q^2 \Pi(q^2) = -i \sum Q_q^2 \text{Tr} \int \frac{\mathrm{d}^4 p_1}{(2\pi)^4} \, \gamma_\mu \frac{1}{\hat{q} + \hat{p}_1 - m_q} \gamma^\mu \frac{1}{\hat{p}_1 - m_q} \ , \quad (3.30)$$

where Tr refers to trace over both, Lorentz and color indices.

To extract the contributions of regions *ii*) and *iii*), we assume that $p_1 \sim m_q \ll q$ or $p_1 - q \sim m_q \ll p_1$. Focusing on the first region, we expand to first order in p_1 and m_q and obtain

$$\Pi^{(ii)}(q^2) = -\sum \frac{Q_q^2}{3q^2} \mathrm{Tr} \int_{p_1 < \mu} \frac{\mathrm{d}^4 p_1}{(2\pi)^4} \left\{ \gamma_\mu \frac{1}{\hat{q}} \gamma^\mu - \gamma_\mu \frac{1}{\hat{q}} (\hat{p}_1 - m_q) \frac{1}{\hat{q}} \gamma^\mu \right\} \frac{i}{\hat{p}_1 - m_q} \,.$$
$$(3.31)$$

When the contribution of region *iii*) is added, the first term in the integrand cancels out and the second term gets multiplied by a factor two. We obtain

$$\Pi^{\mathrm{soft}}(q^2) = 2 \sum \frac{Q_q^2}{3q^2} \mathrm{Tr} \int_{p_1 < \mu} \frac{\mathrm{d}^4 p_1}{(2\pi)^4} \gamma_\mu \frac{1}{\hat{q}} (\hat{p}_1 - m_q) \frac{1}{\hat{q}} \gamma^\mu \frac{i}{\hat{p}_1 - m_q} \,. \qquad (3.32)$$

Upon averaging over directions of q, we find

$$\Pi^{\mathrm{soft}}(q^2) = \sum \frac{Q_q^2}{12q^4} \mathrm{Tr} \int_{p_1 < \mu} \frac{\mathrm{d}^4 p_1}{(2\pi)^4} (8\hat{p}_1 - 32 m_q) \frac{i}{\hat{p}_1 - m_q} \qquad (3.33)$$

$$= \sum \frac{Q_q^2}{12q^4} \left(-8 \langle 0|\bar{q} i \hat{D} q|0 \rangle + 32 \langle 0|m_q \bar{q} q|0 \rangle \right) = \frac{2}{q^4} \langle 0| \sum Q_q^2 m_q \bar{q} q|0 \rangle,$$

which confirms the result $\tilde{c}_q = 2$ obtained earlier. We have used

$$i \int \frac{\mathrm{d}^4 p}{(2\pi)^4} \delta^{ik} (\hat{p} - m)^{-1}_{\beta\alpha} e^{-ipx} = \langle 0|T\{q_\alpha^i(x) \bar{q}_\beta^k(0)\}|0 \rangle \,, \qquad (3.34)$$

where $i, k = 1, \ldots, N_c$ are color indices, and also the equation of motion $i\hat{D}q = m_q q$ to get $\langle 0|\bar{q} i \hat{D} q|0 \rangle = m_q \langle 0|\bar{q}q|0 \rangle$ in the second line of (3.33). The limitation $p_1 < \mu$ implies that the operators that appear in the derivation of the OPE are normalized at μ; in other words, only distances larger than $1/\mu$ contribute to their matrix elements. We note that the validity of equations of motion is *imposed* on the calculation of the OPE; this allows us to deal with divergent expressions that appear in (3.33) in the unique way.

The coefficient function of the operator $\mathcal{O}_4^{\mathrm{gl}}$ is computed from diagrams similar to the one shown in Fig. 3.3b. To perform such computation, it is convenient to start with the expression for the quark propagator in the external

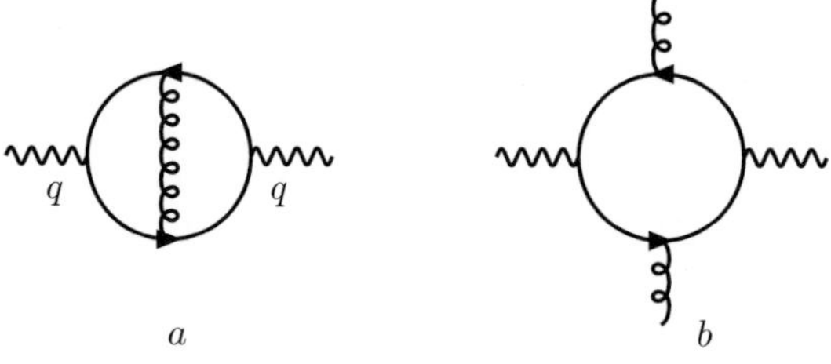

Fig. 3.3. Sample diagrams for computing the operator product expansion of two electromagnetic currents

field and to use the fixed-point gauge $x_\mu A^\mu = 0$ for the vector potential. The details of this approach can be found in [9]. The result of the calculation is $\tilde{c}_{\mathrm{gl}} = \sum Q_q^2/12$. We note that this result can also be obtained from the analysis of momentum flow in two-loop Feynman diagrams shown in Fig. 3.3a. If all the lines in such diagrams are hard, the result is the $\mathcal{O}(\alpha_s)$ correction to the coefficient function of the unity operator c_0. However, if the loop momentum flowing through the gluon line in Fig. 3.3a is soft, the resulting contribution is described in terms of the operator $\mathcal{O}_4^{\mathrm{gl}}$; integration over remaining hard lines in the diagram provides the coefficient function c_{gl}.

For the correlator of the electromagnetic hadronic currents we finally derive

$$\Pi(Q^2) = c_0(Q^2) + \frac{2}{Q^4} \sum Q_q^2 \langle m_q \bar{q}q \rangle$$

$$+ \frac{1}{12Q^4} \sum Q_q^2 \left\langle \frac{\alpha_s}{\pi} G_{\mu\nu}^a G^{\mu\nu,a} \right\rangle + \mathcal{O}(Q^{-6}) . \tag{3.35}$$

The values of the vacuum condensates are determined from QCD sum rules [5]. The summary of most recent results can be found in [10].

We mentioned earlier that $\Pi(Q^2)$ requires renormalization; it contains the ultra-violet divergence. Working with divergent expressions is inconvenient; a simple way to avoid that is to consider derivative of the vacuum polarization function $\Pi(Q^2)$. The corresponding function is the *Adler function* $D(Q^2)$, defined as

$$D(Q^2) = -Q^2 \frac{\partial \Pi(Q^2)}{\partial Q^2} . \tag{3.36}$$

Using (3.35), we derive the operator product expansion for the Adler function

$$D(Q^2) = \frac{N_c}{12\pi^2} \sum Q_q^2 + \frac{4}{Q^4} \sum Q_q^2 \langle m_q \bar{q}q \rangle$$

$$+ \frac{1}{6Q^4} \sum Q_q^2 \left\langle \frac{\alpha_s}{\pi} G_{\mu\nu}^a G^{\mu\nu,a} \right\rangle + \mathcal{O}(Q^{-6}) . \tag{3.37}$$

Equation (3.37) is important because it provides a useful constraint on the e^+e^- hadronic annihilation cross section. To see this, we use the dispersion representation for the photon vacuum polarization $\Pi(Q^2)$, discussed in Sect. 2.4, and derive the following representation for the Adler function

$$D(Q^2) = \frac{1}{12\pi^2} \int\limits_0^\infty ds \, \frac{Q^2}{(s + Q^2)^2} R^{\mathrm{hadr}}(s) . \tag{3.38}$$

In Sect. 3.3, we employ (3.37, 3.38) to constrain a simple model for the e^+e^- hadronic annihilation cross section that we use to estimate the hadronic vacuum polarization contribution to the muon anomalous magnetic moment. Before that, however, we discuss a simple example where the ideas of the large-N_c QCD and the operator product expansion are used to estimate the mass difference of the charged and neutral pions.

3.2.3 An Example: The Charged-neutral Pion Mass Difference

The pions, the lightest mesons in QCD, are the Goldstone bosons that originate from the spontaneous breaking of the $\mathrm{SU}(2)_\mathrm{L} \times \mathrm{SU}(2)_\mathrm{R}$ chiral symmetry to $\mathrm{SU}(2)_\mathrm{L+R}$. Because true Goldstone bosons are massless, pion masses are the consequence of the explicit symmetry breaking in the Lagrangian. The mass quark terms $\sum m_q \bar{q} q$ in the QCD Lagrangian break the $\mathrm{SU}(2)_\mathrm{L} \times \mathrm{SU}(2)_\mathrm{R}$ symmetry and lead to the Gell-Mann–Oakes–Renner relation between the mass of the pion, the quark masses and the quark condensate [11]

$$m_\pi^2 = -\frac{m_u + m_d}{2F_\pi^2} \left\langle \bar{u}u + \bar{d}d \right\rangle , \tag{3.39}$$

where $F_\pi \approx 92$ MeV is the pion decay constant, $\langle 0|\bar{d}\gamma_\mu \gamma_5 u|\pi_p^+\rangle = i\sqrt{2}F_\pi p_\mu$. This relation predicts equal masses for charged and neutral pions; in reality, however, the masses are slightly different

$$m_{\pi^\pm} - m_{\pi^0} \approx 4.6 \text{ MeV} . \tag{3.40}$$

The mass difference of the charged and neutral pions can be understood [12] by invoking electromagnetic interactions which is yet another source of explicit $\mathrm{SU}(2)_\mathrm{L} \times \mathrm{SU}(2)_\mathrm{R}$ symmetry violation in the QCD Lagrangian. In the context of large-N_c QCD the charge-neutral pion mass difference is discussed in [13].

Using Dashen's formula for the masses of the pseudo-Goldstone bosons [14] and the effective Hamiltonian in the one-photon exchange approximation, one can show that the $\pi^\pm$-π^0 mass difference is given by

$$m_{\pi^\pm}^2 - m_{\pi^0}^2 = \frac{e^2}{2F_\pi^2} \int \frac{\mathrm{d}^4 q}{(2\pi)^4} \frac{g^{\mu\nu}}{q^2} \Pi_{\mu\nu}^{V-A} , \tag{3.41}$$

where $\Pi_{\mu\nu}^{V-A}$ is the difference of correlators of the vector and axial currents $V_\mu^- = \bar{d}\gamma_\mu u$ and $A_\mu^- = \bar{d}\gamma_\mu \gamma_5 u$

$$\Pi_{\mu\nu}^{V-A} = \int \mathrm{d}^4 x\, e^{iqx} \langle 0|T\left\{V_\mu^-(x)V_\nu^+(0)\right\} - T\left\{A_\mu^-(x)A_\nu^+(0)\right\}|0\rangle . \tag{3.42}$$

Since we work through first order in the electromagnetic coupling α and zeroth order in the chiral symmetry breaking due to quark mass terms, we need to know the correlators of currents in (3.42) in the limit when the chiral symmetry is exact and the electromagnetic interactions are switched off.

We now discuss constraints on the correlator $\Pi_{\mu\nu}^{V-A}$. First, since in the limit of the exact chiral symmetry the vector and axial currents are conserved, the correlator $\Pi_{\mu\nu}^{V-A}$ is transversal

$$\Pi_{\mu\nu}^{V-A} = i\,(q_\mu q_\nu - g_{\mu\nu}q^2)\Pi^{V-A}(q^2) . \tag{3.43}$$

To find constraints on Π^{V-A} at large q^2, we use the results of the previous section. Then, it is easy to see that, in the chiral limit, the difference between vector and axial current correlators appears *only* at $1/q^6$. This difference is determined by a four-fermion operator that originates from an exchange of a single hard gluon. We write the large-q^2 constraint as

$$\Pi^{V-A}(q^2) \; \xrightarrow{Q^2 \gg \Lambda^2_{\mathrm{QCD}}} \; \alpha_s \, \frac{\langle \bar{\psi}\Gamma\psi\bar{\psi}\Gamma\psi \rangle}{q^6} \; . \tag{3.44}$$

Another constraint follows from the fact that spontaneous breaking of the chiral symmetry leads to the appearance of the massless states in the spectrum of the theory. These Goldstone states are identified with the triplet of pions. It is then easy to see that the correlator Π^{V-A} receives a contribution from a one-pion intermediate state that is singular in the limit $q^2 \to 0$

$$\Pi^{V-A} \; \xrightarrow{q^2 \to 0} \; -\frac{2F_\pi^2}{q^2} \; . \tag{3.45}$$

We may now invoke the large-N_c discussion presented earlier. The mass difference of the charged and neutral pions is N_c-independent, in the limit $N_c \to \infty$; this is consistent with $F_\pi \sim \sqrt{N_c}$ and $\Pi^{V-A}_{\mu\nu} \sim N_c$. In the large-$N_c$ limit any Green's function is represented as an infinite sum over resonances; hence, we write

$$\Pi^{V-A}(q^2) = \sum_V \frac{2g_V^2}{q^2 - m_V^2} - \sum_A \frac{2g_A^2}{q^2 - m_A^2} - \frac{2F_\pi^2}{q^2} \; . \tag{3.46}$$

In (3.46) the summation extends over vector V^i and axial-vector A^i resonances whose couplings are defined as

$$\langle 0|V_\mu^-|V_i\rangle = \sqrt{2}\, m_{V_i}\, g_{V_i} \epsilon_\mu^{V_i} \; , \qquad \langle 0|A_\mu^-|A_i\rangle = \sqrt{2}\, m_{A_i}\, g_{A_i} \epsilon_\mu^{A_i} \; . \tag{3.47}$$

The last term in (3.46) reflects the contribution of massless pions. We require that the large-N_c QCD Anzats for Π^{V-A}, (3.46), complies with the QCD constraint, (3.44). Clearly, (3.46) automatically fulfills the low-q^2 constraint (3.45), but the large-q^2 constraint imposes non-trivial relations between various input parameters.

The large-q^2 constraints are obtained by expanding (3.46) in series of $1/q^2$ and equating the resulting series to the large-q^2 asymptotics in (3.44). From the absence of $1/q^2$ and $1/q^4$ terms in (3.44), we derive two equations

$$\sum_V g_V^2 - \sum_A g_A^2 = F_\pi^2 \; , \qquad \sum_V g_V^2 m_V^2 - \sum_A g_A^2 m_A^2 = 0 \; , \tag{3.48}$$

which are the Weinberg sum rules [15]. These equations constrain the couplings $g_{V,A}$; if we truncate the infinite series in (3.46) to include one vector $\rho(770)$ and one axial vector $a_1(1260)$ mesons, we can solve (3.48) for $g_{V,A}$

explicitely. Assuming that such truncation is a reasonable approximation, we get for the correlator

$$\Pi^{V-A}(q^2) = -\frac{2F_\pi^2 m_V^2 m_A^2}{q^2(q^2 - m_V^2)(q^2 - m_A^2)} \; .$$

(3.49)

Substituting this Π^{V-A} in (3.41), we obtain

$$m_{\pi^\pm}^2 - m_{\pi^0}^2 = \frac{3}{4}\left(\frac{\alpha}{\pi}\right)\frac{m_A^2 m_V^2}{m_A^2 - m_V^2}\ln\frac{m_A^2}{m_V^2} \; ,$$

(3.50)

which translates into a 5 MeV mass difference, in good agreement with the experimental value of 4.6 MeV.

While the agreement with the experimental value is gratifying, we note that the truncation of the infinite sum of resonances to just three terms is non-parametric and is *not guaranteed to work*. For example, the coupling g_ρ computed in the three-meson model, leads to the partial decay width $\Gamma(\rho \to e^+e^-) = 4$ keV, which considerably deviates from the experimental value of 7 keV.

Another way to test the validity of the three-meson model is to observe that (3.49) *predicts* the large-q^2 asymptotic of $\Pi^{V-A}(q^2)$

$$\Pi^{V-A}(q^2) \xrightarrow{Q^2 \gg \Lambda_{\rm QCD}^2} -\frac{2F_\pi^2 m_V^2 m_A^2}{q^6} \; ,$$

(3.51)

which, if the model is viable, should agree with the OPE constraint, (3.44). We will not pursue this analysis further for the charged-neutral pion mass difference, but we note that similar tests of the validity of the large-N_c models are considered in Chap. 5, where the influence of strong interactions on weak corrections to the muon magnetic anomaly is discussed. Finally, we emphasize that a combination of the large-N_c arguments with the OPE constraints at small distances, is a powerful tool to compute Green's functions in QCD. We use this tool in the following section to discuss strong interaction effects in the physics of the muon magnetic anomaly.

3.3 Hadronic Vacuum Polarization: The Theoretical Estimate

We can refine the crude estimate of the hadronic vacuum polarization contribution to the muon anomalous magnetic moment, (3.1, 3.2) by modeling essential features of the hadronic final states produced in the e^+e^- annihilation [16]. To this end, we use the fact that only light hadronic states contribute significantly to $a_\mu^{\rm hvp}$ and construct a model for the e^+e^- annihilation into hadrons that captures essential features of this process for energies below 1 GeV; the rest is described in a crude approximation.

Specifically, for energies smaller than 1 GeV the model includes *i*) the threshold production of two pions — the contribution potentially enhanced by the small pion mass $m_{\pi^\pm} = 139.57$ MeV and *ii*) the production of vector mesons, i.e. the ρ-meson with the mass $m_\rho = 776$ MeV, the ω-meson with $m_\omega = 782$ MeV and the ϕ-meson with $m_\phi = 1020$ MeV. For energies above m_ϕ, we use perturbative QCD to estimate the annihilation cross section.

Since the integrand in (3.3) decreases rapidly for large values of s, we expect that it is the two-pion threshold production that delivers the largest contribution. This expectation is based on the parametrical enhancement of the threshold contribution, $a_\mu \sim (\alpha/\pi)^2 m^2/m_\pi^2$, in the chiral limit when $m_\pi^2 \to 0$. Numerically, however, this expectation turns out to be incorrect and the largest contribution to a_μ^{hvp} comes from the ρ-meson. This can be explained by recognizing that the ρ-meson contribution is enhanced by another "large" parameter in QCD – the number of colors $N_c = 3$. Note, that in an imaginable world with pion and muon masses ten times smaller than their actual values, the chirally enhanced threshold production of the two pions would certainly prevail over all other contributions to a_μ^{hvp}.

Consider first the two-pion contribution to the muon magnetic anomaly. The cross section $\sigma(e^+e^- \to \pi^+\pi^-)$ computed in scalar QED reads

$$\sigma_{e^+e^- \to \pi^+\pi^-} = \frac{\pi\alpha^2}{3s}\left(1 - \frac{4m_\pi^2}{s}\right)^{3/2}. \tag{3.52}$$

This cross section is a good approximation to the actual two-pion production cross section close to the two-pion threshold. When s approaches the mass of the ρ meson, the cross section starts to deviate from (3.52). To avoid double counting, we restrict the integration over the invariant masses in (3.3) by $s_{\max} = m_\rho^2/2$; Substituting (3.52) into (3.3) and integrating over s in the interval $4m_\pi^2 < s < m_\rho^2/2$, we obtain

$$a_\mu^{\mathrm{hvp},2\pi} \approx 400 \times 10^{-11}. \tag{3.53}$$

The contribution of vector mesons to a_μ^{hvp} is computed in the narrow width approximation. This is not an accurate approximation for the ρ-meson whose width is, approximately, one fifth of its mass, but for our purposes it is sufficient. The production cross section of a narrow spin-one resonance V in the e^+e^- annihilation reads

$$\sigma_{e^+e^- \to V}(s) = \frac{12\pi^2 \Gamma_{V \to e^+e^-}}{m_V}\,\delta(s - m_V^2)\,, \tag{3.54}$$

where $\Gamma(V \to e^+e^-)$ is the leptonic partial decay width of the vector meson V. We use $\Gamma(\rho \to e^+e^-) = 7.02$ keV, $\Gamma(\omega \to e^+e^-) = 0.60$ keV and $\Gamma(\phi \to e^+e^-) = 1.27$ keV. Substituting (3.54) into (3.3) and integrating over s, we derive

$$a_\mu^{\text{hvp,vect}} = \sum_{V=\rho,\omega,\phi} \frac{3\Gamma_{V\to e^+e^-}}{\alpha m_V} a_\mu^{(1)}(m_V^2)$$

$$\approx (4732|_\rho + 395|_\omega + 387|_\phi) \times 10^{-11} = 5514 \times 10^{-11}. \quad (3.55)$$

Finally, we approximate the remainder of the hadronic spectrum using perturbative QCD with three massless flavors. The e^+e^- annihilation cross section in this case reads

$$\sigma_{\text{hadr}} = \frac{4\pi\alpha^2}{3s} N_c \sum_{q=u,d,s} Q_q^2 = \frac{8\pi\alpha^2}{3s}, \quad (3.56)$$

where we use $Q_u = 2/3$, $Q_d = Q_s = -1/3$. Substituting (3.56) into (3.3) and integrating over s from 1 GeV^2 to infinity, we obtain

$$a_\mu^{\text{hvp,cont}} = 1240 \times 10^{-11}. \quad (3.57)$$

Adding the three contributions together, we arrive at the estimate

$$a_\mu^{\text{hvp,est}} = a_\mu^{\text{hvp},2\pi} + a_\mu^{\text{hvp,vect}} + a_\mu^{\text{hvp,cont}} = 7160 \times 10^{-11}. \quad (3.58)$$

What are the implications of the above calculation? First, we note that for such a crude estimate, the result (3.58) is astonishingly accurate. A typical recent value for a_μ^{hvp}, obtained by processing large amount of data on the e^+e^- annihilation into hadrons, is [17]

$$a_\mu^{\text{hvp}} = 6963(80) \times 10^{-11}; \quad (3.59)$$

it agrees with the estimate (3.58) to within three percent! Clearly, the degree of agreement between the two results is accidental; however, it shows that relatively few features of the hadronic spectrum in the e^+e^- annihilation determine the bulk of a_μ^{hvp}.

Another important lesson from the above computation is that it is difficult to estimate the uncertainty that has to be assigned to it. In particular, the choice of $m_\rho^2/2$ as the upper integration limit for the two-pion contribution and the choice of 1 GeV as the lower integration limit for the continuum contribution are arbitrary. In both cases there is a strong sensitivity to the cut-off, so that our estimate (3.58) can change by ~ 10 percent. Unfortunately, this is a common problem of model-based calculations – even when central values that are derived are reasonable, accurate uncertainty estimates are difficult.

In spite of that, there is an instructive way to determine the proper value of the lower limit of integration in the integral over continuum contribution, s_0; we will use a similar approach in a forthcoming discussion of electroweak and hadronic light-by-light scattering contributions to the muon anomalous magnetic moment. The constraint on s_0 comes from the fact that there is an additional requirement for our model of the e^+e^- hadronic annihilation cross section that we failed to utilize so far. It comes from the short-distance properties of QCD. Using our model for σ_{hadr}, we can compute the large-Q^2 asymptotics of the photon vacuum polarization and compare it with the results based on the operator product expansion, discussed in Sect.3.2.2. For such a comparison, it is convenient to work with the Adler function (3.36).

We use our model for $\sigma_{\text{hadr}}(s)$ to compute the Adler function $D(Q^2)$, in the limit $Q^2 \to \infty$. In doing so, we neglect the pion threshold production which turns out to be small. We derive

$$12\pi^2 D^{\text{mod}} = 2 + \frac{1}{Q^2}\left(\sum_{V=\rho,\omega,\phi} \frac{9\pi \Gamma_{V\to e^+e^-} m_V}{\alpha^2} - 2s_0 \right) + \mathcal{O}(Q^{-4}) \; . \quad (3.60)$$

There is a striking difference between the two results for the Adler function (3.60) and (3.37). The OPE-based calculation predicts[2] the *absence* of the $1/Q^2$ term in the Adler function whereas our model (3.60) fails to agree with this result, unless we identify

$$s_0 = \sum_{V=\rho,\omega,\phi} \frac{9\pi \Gamma_{V\to e^+e^-} m_V}{2\alpha^2} \approx (1.38 \text{ GeV})^2 \; . \quad (3.61)$$

With this choice of the parameter s_0, our estimate of the hadronic vacuum polarization contribution becomes

$$a_\mu^{\text{hvp,est}} \approx 6571 \times 10^{-11} \; . \quad (3.62)$$

The difference between the estimate (3.62) and the data-based calculations is about six percent. We emphasize that, although this is too large a difference to make the theoretical computation of the hadronic vacuum polarization to be of practical use for a_μ phenomenology, it is remarkably accurate for the quantity that, almost entirely, is determined by non-perturbative dynamics.

An important lesson to draw from the above discussion concerns the hierarchy of different contributions to a_μ^{hvp}. We have found that

$$a_\mu^{\text{hvp,2}\pi} \lesssim a_\mu^{\text{hvp,cont}} \ll a_\mu^{\text{hvp, vect}} \; , \quad (3.63)$$

which implies that the chirally enhanced contribution – the two-pion threshold production – is, in fact, the smallest. We therefore see that the smallness of the pion mass and, consequently, the chiral enhancement, is of little relevance for a_μ^{hvp}. On the other hand, large contributions due to the ρ-meson and the continuum can be understood since they are enhanced by another important QCD parameter, the number of colors N_c. We will see a very similar pattern of chirally-enhanced vs. N_c-enhanced contributions in Chap. 6 where the hadronic light-by-light scattering component of the muon magnetic

---

[2] See, however, a warning in the footnote on page 42.

anomaly is discussed. In that case, there is no data to cross-check our understanding of the low-energy hadron physics; hence, the intuition developed in connection with the hadronic vacuum polarization will be of great help there.

Before discussing the data-driven evaluation of the hadronic vacuum polarization component of the muon anomalous magnetic moment, we mention that, recently, the first lattice calculation of a_μ^{hvp} was undertaken [18]. While the current result of the quenched calculation $a_\mu^{\mathrm{hvp}} = 4600(780) \times 10^{-11}$ is far from high precision required for this contribution, it is pointed out in [18] that the improvement in precision is possible. Apart from obvious lattice issues, such as finite lattice spacing and finite volume effects, the quenched approximation has to be given up entirely to arrive at a meaningful result. The attempts to compute a_μ^{hvp} beyond the quenched approximation are currently underway [19].

The theoretical estimate of the hadronic vacuum polarization contribution to a_μ discussed in this section shows that it is quite significant numerically. Because of that, its precise evaluation is required. In the next section we discuss basics of the data-driven analysis that is the primary source of our knowledge of the hadronic vacuum polarization contribution to a_μ.

3.4 Basics of the Data-driven Analysis

3.4.1 Preliminary Remarks

As we explained in the previous sections, computation of a_μ^{hvp} is contingent upon the knowledge of the e^+e^- annihilation cross section into hadrons. Therefore, making theoretical predictions for a_μ^{hvp} requires dealing with experimental data. Experiments at e^+e^- colliders started in late sixties and continue ever since; over this time, a tremendous wealth of data was accumulated. These data are of different quality; later experiments have access to better technology and build upon the experience of their predecessors. Since new experimental data are continuously produced, theoretical predictions for the hadronic vacuum polarization contribution to a_μ evolve in time (see Table 3.1). In general, later analyses have access to more recent and better experimental data and, for this reason, naturally supersede the older ones.

A glance at Table 3.1 reveals that the current precision on hadronic vacuum polarization contribution to a_μ is higher by a factor $2 - 3$, compared to the analyses twenty years ago [20, 21]. Significant part of the improvement comes from new data on the e^+e^- annihilation; in addition, various theoretical ideas on how to increase the precision of a_μ^{hvp} are being explored. These ideas include an aggressive use of perturbative QCD at fairly low energies as well as incorporating data on hadronic decays of τ-leptons, corrected for isospin violations, into the computation of a_μ^{hvp}. While the use of these ideas became wide-spread in recent years, virtues of such approaches are questionable. Substituting experimental data by theoretical analysis may lead to an illusion of the error reduction; however, in the long run such improvements

Table 3.1. Computations of the hadronic vacuum polarization contribution to a_μ since 1985. As indicated, some of the analyses use data on hadronic τ-decays or partially rely on perturbative QCD

Ref.	$a_\mu^{\mathrm{hvp}} \times 10^{11}$	Comments
Ref.[20]	6840(110)	e^+e^-
Ref.[21]	7070(180)	e^+e^-
Ref.[22]	7100(115)	QCD and e^+e^-
Ref.[23]	7048(115)	QCD and e^+e^-
Ref.[24]	7024(153)	e^+e^-
Ref.[25]	7026(160)	e^+e^-
Ref.[26]	6950(150)	e^+e^-
Ref.[26]	7011(94)	e^+e^- and τ
Ref.[27]	6951(75)	e^+e^-, τ and pQCD
Ref.[28]	6924(62)	e^+e^-, τ, pQCD and sum rules
Ref.[17]	6963(71)	e^+e^-
Ref.[17]	7110(57)	τ
Ref.[1]	6924(64)	e^+e^-
Ref.[29]	6934(63)	e^+e^-, incl. KLOE
Ref.[30]	6948(86)	e^+e^-
Ref.[31]	6944(49)	e^+e^-

usually result in a controversy which requires new experimental input to sort things out.

In the remainder of this section various aspects of the data-driven calculations of a_μ^{hvp} are summarized. Before we dive into this discussion, a few general remarks are appropriate. We pointed out in the previous section that only relatively light hadronic states substantially contribute to a_μ^{hvp}. This is illustrated in Tables 3.2 and 3.3, where full hadronic contribution to a_μ^{hvp} is split by channels and by energy regions. The results given in those Tables are approximate and do not always incorporate the most up to date data.

Table 3.2. Contributions of various channels to a_μ^{hvp}. The last column refers to the precision with which a particular channel is claimed to be known. The results given in the Table are approximate and do not always incorporate the most up to date data

channel	$a_\mu^{\mathrm{hvp}} \times 10^{11}$	Precision achieved
$\pi^+\pi^-$	5060	$\leq 1\%$
$\omega \to 3\pi$	380	3%
ϕ	357	5%
4π	310	5%
3π	66	10%
K^+K^-	46	10%
$2\,\mathrm{GeV} \leq \sqrt{s} \leq 5\,\mathrm{GeV}$	200	5%
$\sqrt{s} \geq M_\Upsilon$	40	$\leq 1\%$

It is apparent from Table 3.2 that the largest contribution to a_μ^{hvp} comes from the two-pion channel which, in turn, is dominated by the ρ-meson resonance. Because of that, the two-pion production in the e^+e^- annihilation has to be known with very high precision. Since other hadronic channels contribute much less than the $e^+e^- \to \pi^+\pi^-$, the precision requirements for them are more modest.

From Table 3.3 it follows that approximately ninety percent of the hadronic vacuum polarization contribution to a_μ comes from low energies, $\sqrt{s} \leq 2\,\mathrm{GeV}$. Since $m_\tau \approx 1.8\,\mathrm{GeV}$, the most important channels such as $2\pi, 4\pi$, etc. can be studied in hadronic decays of the τ lepton. As we explain below, this is non-trivial since, with the current precision on a_μ^{hvp}, theoretical corrections have to be applied to data on hadronic τ decays for the purpose of a_μ^{hvp} evaluation.

Important issues that have to be discussed in connection with the data-driven evaluations of a_μ^{hvp} include i) validity of theoretical improvements suggested in recent years; ii) QED radiative corrections; iii) combination of data from different experiments; iv) contributions to a_μ^{hvp} from different energy regions; v) a special role of the $e^+e^- \to \pi^+\pi^-$ channel; vi) consistency of different data-driven analyses. We describe those issues in the remainder of this section.

Table 3.3. Contributions of various energy regions to a_μ^{hvp}, from [1]

Energy region, GeV	$a_\mu^{\mathrm{hvp}} \times 10^{11}$	$\delta a_\mu^{\mathrm{hvp}} \times 10^{11}$
$2m_\pi - 1.4$	6089	± 52
1.4-2.0	319	± 24
2.0 -11	420	± 11.4
J/ψ	73	± 4
Υ	1.09	± 0
$11 - \infty$	21	± 0
Total	6924	± 59

3.4.2 Theoretical Issues

There are some theoretical aspects of the data-driven analysis that became important in recent years. They can be divided into two broad categories: *i) theoretical* suggestions on how the precision of a_μ^{hvp} can be further improved and *ii)* issues that have to be addressed because the precision of a_μ^{hvp} evaluations has increased greatly.

An example of the first kind is an aggressive use of the operator product expansion down to fairly low energies, pioneered in [22], and used recently in [1, 27] to estimate contributions of energy regions where the quality of the data is poor. As we explained in Sect. 3.2, the operator product expansion in its standard form is based on the *assumption* that non-perturbative contributions to the coefficient functions of the operators could be neglected. While this assumption is plausible, it can not be justified from first principles; hence, the operator product expansion can not be used as a substitute for the experimental data.

Yet another example of a similar sort is the suggestion by Alemany, Davier and Höcker [26] to use data on hadronic τ decays as a complimentary source of information about hadronic vacuum polarization. While [26] claimed a substantial reduction of the theoretical uncertainty in a_μ^{hvp} as the result of including the τ data, this issue was marred almost from the very beginning. Immediately after the suggestion of [26], a more precise data on the e^+e^- annihilation into hadrons [32] showed systematic differences between the e^+e^- and the τ data [33]. While, originally, the problem was not statistically significant, further experimental studies [34, 35] only reinforced the conclusion that there is a conflict between the e^+e^- and the τ data. At the moment, the difference in the hadronic vacuum polarization contributions to the muon anomalous magnetic moment evaluated with the e^+e^- data and the τ data is, approximately, two standard deviations.

Currently, there is no resolution of the $e^+e^- - \tau$ puzzle. The experimental data on both the e^+e^- annihilation and the hadronic τ decays are confirmed by more than one experiment [34, 35, 36, 37, 38, 39], so it seems that the experimental origin of the problem may be excluded. However, such conclusion is somewhat premature. Indeed, while all recent e^+e^- experiments have comparable precision, the situation with the data on τ decays is different. In particular, large disagreement between the e^+e^- and the τ data occurs because of a very high precision claimed by the ALEPH collaboration [36]; the two other results on hadronic τ decays by CLEO and OPAL have larger errors and, hence, are much less inconsistent with the e^+e^- data.

In addition, there is an important conceptual difference between the e^+e^- and the τ data for the purpose of the a_μ^{hvp} evaluation. While the e^+e^- hadronic annihilation cross section directly enters the dispersion integral (3.4), the τ data can be used there only after the isospin symmetry breaking corrections are applied. As we discuss in Sects. 3.5, 3.6, computation of such correction is non-trivial and brings in significant theoretical input into, otherwise, experimentally driven analysis. Because of potential caveats with using the τ data and since the e^+e^- data is on the solid footing, the use of the τ data for a_μ^{hvp} computation is somewhat downplayed in most recent analyses [29].

As we mentioned earlier, since the precision on a_μ^{hvp} is high, new issues become important. An example is provided by the QED radiative corrections to the reaction $e^+e^- \to X_{\mathrm{h}}$, where X_{h} is a hadronic state. Although such corrections are proportional to $\alpha/\pi \sim 10^{-3}$ and for this reason may be considered irrelevant, they are usually dynamically enhanced and, hence, are quite important for the purpose of a_μ^{hvp} computation. In particular, the photon radiation from the initial state in $e^+e^- \to X_h$ is enhanced by a logarithm $\ln(\sqrt{s}/m_e)$ which, for $\sqrt{s} \sim m_\rho \sim 770$ MeV is equal to seven, approximately. Another example concerns the photon radiation off charged particles in the final state X_h; experimental constraints applied to identify particular hadronic state usually cut-off radiation of photons at moderate $\gtrsim 10°$ angles relative to momenta of charged hadrons. Such constraints also lead to a logarithmic enhancement of the QED radiative corrections making them of the order of a few percent. There are other sources of dynamical enhancement of the QED radiative corrections; we will not discuss them here. However, the bottom line is that since we require a_μ^{hvp} with the precision of about one percent, it is important to understand which QED corrections have to be included in what is called σ_{hadr} in (3.4) and how this σ_{hadr} is related to a quantity that is published as the e^+e^- hadronic annihilation cross section.

The QED radiative corrections to $e^+e^- \to X_{\mathrm{h}}$ can be decomposed into four pieces – corrections to the initial state, corrections to the final state, the vacuum polarization correction and the interference effects between the initial and final state.[3] From the derivation of the convolution integral (3.3), it is quite obvious that σ_{hadr} that enters (3.4) is defined to include the QED corrections to the final state *only*. On the other hand, the measured annihilation cross section of $e^+e^- \to X_h$ usually includes the initial state radiation, vacuum polarization correction and *only partially* final state QED radiative corrections since experimental cuts, as a rule, reject the emission of hard, non-collinear photons off the final state particles. Hence, what is needed theoretically in (3.3) is almost exactly the opposite to what is being measured experimentally, when the QED effects are taken into consideration.

To arrive at the quantity needed to compute a_μ^{hvp}, one has to remove the initial state radiation and the vacuum polarization QED effects from the measured e^+e^- annihilation cross sections. In addition, the luminosity, traditionally measured by considering standard QED processes such as $e^+e^- \to \mu^+\mu^-$ and large/small angle Bhabha scattering, must be determined with the adequate precision as well. While these issues were largely ignored in earlier calculations of a_μ^{hvp}, the two latest computations [1, 17] include the correction factors to account for QED effects if such effects are not considered in original publications.

3.4.3 Combining the Data

As a matter of principle, given experimental data on the e^+e^- hadronic annihilation cross section, it is straightforward to compute the convolution integral (3.3) and obtain the hadronic vacuum polarization contribution to the muon anomalous magnetic moment. However, in practice, it is a highly nontrivial enterprize. Since data from different experiments has to be combined, the estimate of the uncertainty should be carefully addressed.

Traditionally, data on the e^+e^- annihilation to hadrons is published for exclusive channels for center of mass energies $\sqrt{s} \leq 2$ GeV, whereas the inclusive cross section is measured at higher energies. In the region $1.4\,\mathrm{GeV} \leq \sqrt{s} < 2$ GeV there are both, exclusive and inclusive data which may be used for the analysis.

There are two possible ways to proceed with the computation of the convolution integral (3.3). One may take the results of different channels/experiments separately, integrate them over s and then combine the results of such integrations to obtain a_μ^{hvp}. Unfortunately, this ignores the issue of errors, correlated between different experiments and is also not practical for data sets with small number of points.

The second way to obtain the e^+e^- hadronic annihilation cross section for $\sqrt{s} \leq 2$ GeV is to combine data on a given hadronic channel from different experiments and integrate over energy. The estimate of the error on

---

[3] When total cross sections are considered, the interference of photon radiation from initial and final states at order $\mathcal{O}(\alpha)$ vanishes because of the charge conjugation symmetry. The interference effects may appear in higher orders in QED but they are too small to be of any practical relevance.

the final result becomes a crucial issue. First, there are data sets that are not compatible, within the quoted errors. Second, systematic errors in different experiments and/or different channels are often correlated. For example, all experiments might have a common systematic error related to the lack of knowledge of higher order QED radiative corrections that affects the luminosity measurements. Similarly, for multi-hadronic channels, computation of experimental acceptances relies on Monte-Carlo event generators that are based on common dynamical models.

The fit to data proceeds in the following way [17, 26]. First, a particular hadronic channel and the energy range is chosen. The energy interval is then split into N_b bins. The best value of the cross section, for each bin, is found by minimizing the function

$$\chi^2 = \sum_{n=1}^{N_{\text{exp}}} \sum_{i=1}^{N_b} (\sigma_i^n - \bar{\sigma}_i)_i \, C_{n,ij}^{-1} \left(\sigma_j^n - \bar{\sigma}_j\right) \, , \tag{3.64}$$

where σ_i^n is the cross section measured in experiment n for the value of energy that falls into i-th bin. $\bar{\sigma}_i$ is the cross section averaged between different experiments. $C_{n,ij}$ is the covariance matrix between two energy bins in the experiment n

$$C_{n,ij} = \begin{cases} \left(\delta_{i,\text{stat}}^n\right)^2 + \left(\delta_{i,\text{sys}}^n\right)^2, & \text{for } i = j\,; \\ \delta_{i,\text{sys}}^n \delta_{j,\text{sys}}^n\,, & \text{for } i \neq j\,. \end{cases} \tag{3.65}$$

Minimizing (3.64) with respect to N_b variables $\bar{\sigma}_i$, we obtain the average cross section. The inverse covariance matrix between the solutions $\bar{\sigma}_{i,j}$ is given by

$$\tilde{C}_{ij}^{-1} = \sum_{n=1}^{N_{\text{exp}}} C_{n,ij}^{-1} \, . \tag{3.66}$$

This covariance matrix is used to calculate the uncertainty of a_μ^{hvp} when the dispersion integral (3.3) is computed.

Unfortunately, this procedure may lead to a situation when χ^2 per degree of freedom is larger than one; this signals that experimental data used in the computation are mutually inconsistent. When this happens, errors reported by original experiments are enlarged to allow those data sets to be combined consistently. In [1] a different procedure based on the non-linear χ^2 function is adopted. In addition to $\bar{\sigma}_i$, relative normalization of data points from different experiments is determined from the fit. In this way, the necessity to re-scale the errors is avoided.

Average values of cross sections for a given channel $\bar{\sigma}_i$ and their uncertainties given by the covariance matrix C_{ij} permit the computation of the dispersion integral (3.3) with the help of the trapezoidal rule. While this method is reliable when the function changes slowly and/or the energy bins are sufficiently fine, care has to be exercised when the trapezoidal rule is applied to integrate rapidly changing functions with sparse data points as, for example, is the case when narrow resonances are treated. An alternative, adopted in early studies of the hadronic vacuum polarization contribution to a_μ is to assume a certain functional form of $\sigma_{\mathrm{hadr}}(s)$, determine parameters of the function by fitting the data and then integrate the function over s. While this approach offers certain advantages, it was mostly discarded as too prone to "prejudices" concerning the functional form of $\sigma_{\mathrm{hadr}}(s)$.

3.4.4 Specific Contributions to the Convolution Integral

We now discuss specific contributions to the convolution integral (3.3). The areal view of $R^{\mathrm{hadr}}(s)$, (3.4), is shown in Fig. 3.4. Since the e^+e^- annihilation cross section is a rapidly changing function of the center of mass energy, it is necessary to split the computation of the convolution integral into several

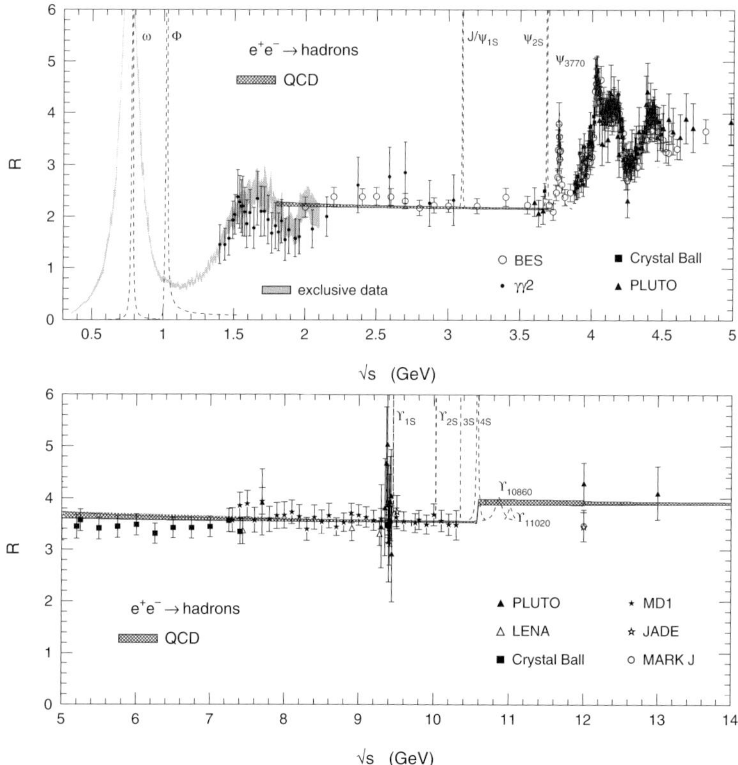

Fig. 3.4. Areal view of $R(s)$. The cross-hatched band gives the prediction from perturbative QCD, which is found to be in good agreement with the measurements in the continuum above 2 GeV [41]

components and to treat each of them separately. The following components are usually identified.

(1) The energy range $\sqrt{s} \geq 11.5$ GeV. In this energy range, we expect the perturbative QCD to be applicable for the evaluation of $\sigma_{\mathrm{hadr}}(s)$. In the approximation that quarks in the final state are massless, σ_{hadr} is known through $\mathcal{O}(\alpha_s^3)$ [42]. If quark masses are retained, the $\mathcal{O}(\alpha_s^2)$ result is available [43]. Those perturbative results can be supplemented by non-perturbative contributions derived in the context of the operator product expansion in QCD, Sect. 3.2.2. The contribution from this energy domain, $\sqrt{s} > 11.5$ GeV, to a_μ^{hvp} is approximately 20×10^{-11} and the error is negligible.

(2) Narrow heavy quark resonances $J/\psi, \psi', \Upsilon$. These narrow resonances require a separate treatment and there are various ways to do that. In [17], the Breit-Wigner approximation is adopted, while [1] treats those resonances in the narrow width parametrization where the production cross section of a resonance of mass M_r in the e^+e^- annihilation is proportional to a delta function $\delta(s - M_r^2)$. While there is a few percent difference in the results of [17] and [1], this discrepancy is irrelevant in practice since the contribution of narrow heavy quark resonances to a_μ^{hvp} is small, $\sim 70 \times 10^{-11}$.

(3) The energy range 2 GeV $\leq \sqrt{s} \leq 11.5$ GeV. In this energy range, the inclusive data on $R(s)$ is used. For a while, there was a problem with the quality of the data in the energy ranges 2 GeV $\leq \sqrt{s} \leq 3.7$ GeV. The situation has improved dramatically thanks to recent BES measurements [44] that are systematically lower and much more precise than the Mark I and $\gamma\gamma 2$ results [45, 46]. It is interesting to note that the first indication that the e^+e^- hadronic annihilation cross section measured by Mark I and $\gamma\gamma 2$ collaborations is too high came from confronting it with perturbative QCD estimates [24, 27].[4]

(4) The energy range 1.4 GeV $\leq \sqrt{s} \leq 2$ GeV. The peculiar feature of this energy region is the fact that *both*, the inclusive measurement of R^{hadr} as well as measurements of individual hadronic channels are available. This offers the possibility to compare the sum over exclusive modes to the inclusive measurement. When this is done, there appears to be a problem since the sum over exclusive modes is *larger* than the inclusive result; in terms of their contribution to a_μ^{hvp}, the difference of the exclusive and inclusive approaches is about 40×10^{-11}. In [1] there is a dedicated discussion of this issue; the authors concluded that the results of the inclusive measurements are more appropriate. We will discuss this issue in Sect. 3.4.6.

(5) The energy range 0.5 GeV $\leq \sqrt{s} \leq 1.5$ GeV: *multi-particle hadronic channels*. In this energy range various exclusive channels are traditionally

---

[4] The Mark I data in the energy range 5 GeV $\leq \sqrt{s} \leq 7$ GeV was also known to be significantly higher than the Crystal Ball data [47] as well as perturbative QCD predictions [24].

measured. For example, there are measurements of the $e^+e^- \to n\pi$ cross sections, up to $n = 6$. There is also data on K^+K^-, $K_S K_L$, $KK\pi$ as well as $2K2\pi$ cross sections. Some of those channels are actually saturated by decays of ω and ϕ mesons. This happens, for example, in case of $\pi^+\pi^-\pi^0$, which receives major contributions from $\omega \to \pi^+\pi^-\pi^0$, with both continuum production and $\phi \to \pi^+\pi^-\pi^0$ playing only a minor role. A similar situation happens for the K^+K^- and $K_S K_L$ channels that are saturated by ϕ decays.

Hadronic channels dominated by resonances can be treated easily since high-quality data is available. On the contrary, the four-pion channels are dominated by the continuum and the quality of the data is poor. The production of five and six pions in the e^+e^- annihilation gives a small, yet non-negligible contribution to a_μ^{hvp}.

(6) The energy range 0.5 GeV $\leq \sqrt{s} \leq 0.9$ GeV: *the two-pion channel.* This energy range gives by far the largest contribution to a_μ^{hvp}. Because of that, $\sigma(e^+e^- \to \pi^+\pi^-)$ has to be known with the precision better than one per-cent. The new data from CMD-2, SND and, to a certain extent, from KLOE satisfy this requirement and allow accurate determination of the two-pion contribution to a_μ^{hvp}. We discuss the two-pion contribution in more details below.

(7) The energy range $2m_\pi \leq \sqrt{s} \leq 0.5$ GeV. Due to the fact that, close to the threshold, the pions are produced in a P-wave, the cross section $\sigma(e^+e^- \to \pi^+\pi^-)$ is proportional to the third power of the relative velocity of the two pions and vanishes rapidly once the threshold is approached. Consequently, there is not much data for $\sqrt{s} \sim 2m_\pi$ and the $e^+e^- \to \pi^+\pi^-$ annihilation cross section in this region has to be estimated from other sources. This is usually done by adopting the polynomial parametrization for the pion form factor inspired by chiral perturbation theory; the parameters are determined from a fit to data on the pion form factor for both time-like and space-like momentum transfers.

3.4.5 The Two-pion Channel

The reaction $e^+e^- \to \pi^+\pi^-$ provides the largest contribution to the hadronic vacuum polarization component of the muon magnetic anomaly and, hence, requires special attention. From Table 3.3 we see that this channel contributes up to seventy percent of a_μ^{hvp}; the major part comes from the production of the ρ resonance that decays into $\pi^+\pi^-$. Because the two-pion production gives such a large contribution to a_μ^{hvp} it has to be computed very precisely; this requires high quality data for $e^+e^- \to \pi^+\pi^-$. Such data recently appeared from three sources; the CMD-2 and SND collaborations measured the pion form factor using traditional energy scan [34, 39], while the KLOE collaboration measured [35] the production of the two pions by the radiative return method [40]. The results of these measurement are in fair agreement with each other *although* some inconsistencies

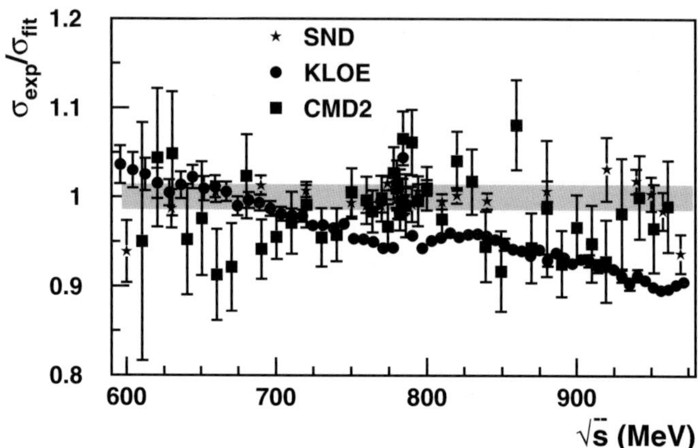

Fig. 3.5. The comparison of the $e^+e^- \to \pi^+\pi^-$ annihilation cross sections measured by CMD-2, SND and KLOE in the energy region around the ρ-resonance. From [39]

are present. In as much as the two-pion contribution to the muon magnetic anomaly is concerned, the results of integrating the experimental data over $0.6\,\mathrm{GeV} < \sqrt{s} < 1\,\mathrm{GeV}$ are $3856(52) \times 10^{-11}(\mathrm{SND})$, $3823(36) \times 10^{-11}(\mathrm{CMD\text{-}2})$ and $3750(50) \times 10^{-11}(\mathrm{KLOE})$. Note that the CMD-2 results were recently updated. A reasonable agreement of the results of the different measurements is overshadowed by the fact that there is a significant difference in the *shape* of the $e^+e^- \to \pi^+\pi^-$ annihilation cross section in the energy region around the ρ peak, Fig. 3.5. However, it is clear a priori that an agreement of spectral densities is a more demanding subject than the agreement of the integrals over spectral densities because such integrals are more stable against various effects that may cause significant disagreement in a point-wise comparison. In particular, the shape difference between the results by KLOE and the direct measurements in e^+e^- annihilation by CMD-2 and SND is reminiscent of the radiative corrections to the production of a narrow resonance.

For the two-pion channel, the correct inclusion of the QED radiative corrections is very important; this concerns *both* the exact definition of the "two-pion" final state and the luminosity measurement. Recall, that the final state radiation off the pions has to be included into the "two-pion" channel, for the purpose of a_μ^{hvp} evaluation. However, traditionally, in e^+e^- experiments the two-pion final state is defined as the two energetic pions whose momenta are back-to-back to an extent specified by an acollinearity cut. This cut therefore *rejects* kinematic configurations when a hard photon is emitted off the pion at a relatively large angle. The rejected contribution is added back *theoretically*, using scalar QED to estimate the photon emission rate at large angles. While the large angle photon emission is not a big effect and so most likely the error introduced in this way is insignificant, it should be kept in mind that this treatment introduces a model-dependence into, otherwise, data-driven analysis. In the long run, there is a need to better understand the range of validity of applying scalar QED to describe the radiative corrections for the two-pion final state. First steps in this direction are taken by the KLOE collaboration that uses the charged pion forward-backward asymmetry, linearly sensitive to the final state radiation amplitude, to investigate potential deviations from the scalar QED model. In addition, theoretical analysis of the two-pion channel is being extended to account for the "pion structure", albeit in the framework of chiral perturbation theory [48].

3.4.6 Comparative Analysis

It should be clear from the above discussion that many details have to be taken into account when the compilation of data and the subsequent evaluation of the dispersion integral (3.3) is attempted. To a large extent, the authors of such analyses have to make choices that are subjective. How does such subjectivity influence the results of the analyses? To answer this question it is instructive to compare two recent evaluations of a_μ^{hvp}, [17] (DEHZ) and [1] (HMNT). From Table 3.1 we see that the difference[5] between the two results is $\sim 40 \times 10^{-11}$ or, approximately, 0.6%. Given the error bars of the two analyses, this should be regarded as the perfect agreement. Nevertheless, it is interesting to compare the two evaluations to understand where this difference comes from.

Surprisingly, almost the entire difference comes from the region $1.4\,\text{GeV} \leq \sqrt{s} \leq 2\,\text{GeV}$, where DEHZ use the sum of exclusive channels to evaluate σ_{hadr} whereas HMNT use the results of the inclusive measurements. Have the sum over exclusive channels been used in HMNT analysis, their result would go up by $\sim 43 \times 10^{-11}$ and would then be in complete agreement with DEHZ.

There are other appreciable differences in the analyses of [17] and [1] that, to a large extent, cancel out when the sum over all contributions to a_μ^{hvp} is computed. Since the cancelation is accidental, it is instructive to discuss the differences. There are essentially two channels where such a disagreement occurs – the threshold production of the two pions and the production of four pions in the reaction $e^+e^- \to \pi^+\pi^-2\pi^0$.

The difference in the two-pion channel occurs, surprisingly, in estimating the small contribution to a_μ from the threshold pion production. Both DEHZ and HMNT use the chiral expansion of the pion form factor and fit the parameters of the expansion using the available data. Although the approaches of the two groups are similar, the results disagree. Part of the reason is that DEHZ do not use data from NA7 experiment [49], claiming that those data suffer from a bias in the energy scale determination. Unfortunately, the NA7 data is quite important since it is the only direct measurement of the two-pion production cross section all the way down to the two-pion threshold.

---

[5] From [17] we take the result based on the e^+e^- data.

The other channel where uncomfortably large disagreement between [17] and [1] exists is the $\pi^+\pi^-2\pi^0$. There, part of the disagreement is also explained by the difference in the input data; in contrast to DEHZ, HMNT include the $\gamma\gamma2$ data [46] into their analysis. It should be pointed out, however, that experimental data on $\pi^-\pi^+2\pi^0$ is not very accurate and significant discrepancies are observed between different experiments. The problems are related to systematic uncertainties in computing a fairly small, $\sim 20-30\%$, acceptance for this channel.

Finally, we return to the discussion of the inclusive vs. exclusive measurements in the region $1.4\ \text{GeV} \leq \sqrt{s} \leq 2\ \text{GeV}$; this is the single major source of the difference between DEHZ and HMNT results. As we mentioned earlier, the measured inclusive e^+e^- annihilation cross section in this energy range is *smaller* than the sum over all exclusive channels.[6] The difference between the inclusive and the exclusive measurements in this energy range is, formally, about 2 standard deviations, with the precision for each of the measurements of the order of five to ten percent. While this level of disagreement is not disastrous, it is quite puzzling.

The ambiguity can be resolved [1] if one invokes the duality argument which implies that suitable quantities averaged over sufficiently broad energy domain can be computed using the operator product expansion in QCD. One may then consider the integral

$$I[f] = \int\limits_{s_1}^{s_2} \mathrm{d}s f(s) R^{\text{hadr}}(s)\,, \tag{3.67}$$

for a sufficiently smooth function $f(s)$. Evaluating (3.67) using the operator product expansion in QCD one gets a theoretical estimate for $I[f]$ which is compared to the data-based evaluation. HMNT showed that the theoretical estimate is much closer to what is obtained when data from *inclusive* measurements is used to evaluate $I[f]$. These perturbative QCD calculations should be considered as a strong evidence that there are potential problems with exclusive measurements in this energy range. Recall, that an earlier application of perturbative QCD in the energy range $2\ \text{GeV} \leq \sqrt{s} \leq 3.7\ \text{GeV}$ in [27] was quite successful; the predicted deficiency of the data from Mark I and $\gamma\gamma2$ in this energy range was later confirmed by recent measurements at BES.

Nevertheless, it is important to keep in mind that only experimental measurements can provide a model-independent determination of a_μ^{hvp} and for this reason we feel that the discrepancy in data should lead to an uncertainty in the final result that is large enough to incorporate both data sets. In this

---

[6] Traditionally, the "inclusive" measurement of the cross section in this energy range is defined as the measurement from where *all the two-particle final states, e.g. $\pi^+\pi^-$, K^+K^-, etc.* are omitted. Those channels have to be added back to arrive at what is usually understood under the inclusive cross section.

regard, the significance of the proximity of the inclusive result to perturbative QCD estimates should not be overstated.

3.4.7 Final Remarks

The comparison between the most recent and up-to-date evaluations of a_μ^{hvp} [1, 17, 29, 30, 31] shows that different data-driven computations essentially converged. Issues that remain problematic merely reflect preferences of the authors of a given analysis, as far as the quality of existing data is concerned. Further improvement in the data-driven evaluations of a_μ^{hvp} is only possible if new data for problematic channels and/or energy regions appear. Potential sources of such new data are B-factories, operating at the center of mass energy $\sqrt{s_b} \sim 11$ GeV. Using the radiative return method [40], it is possible to measure the e^+e^- hadronic annihilation cross section for $\sqrt{s} < \sqrt{s_b}$. Studies at BaBar indicate that this is a viable possibility [50, 51].

3.5 Hadronic Vacuum Polarization: Inclusion of the τ-data

In 1998, Alemany, Davier and Höcker suggested [26] to improve the precision of the hadronic vacuum polarization contribution to the muon anomalous magnetic moment by incorporating high-quality data on hadronic τ-decays. Since then, for a few years, the τ data was the major force driving the increase in precision of a_μ^{hvp}; it was a standard practice to average a_μ^{hvp} computed using the e^+e^- and the τ data and to reduce the uncertainty in that way.

Unfortunately, the inclusion of the τ data is not straightforward theoretically since in τ decays hadrons are produced by a charged current. To relate these data to the hadroproduction by an electromagnetic current, needed for a_μ^{hvp} calculation, we use the isospin symmetry. Since this symmetry is not exact, we need to account for isospin violations. It turns out that the isospin violation corrections are difficult to compute since they are sensitive to hadron dynamics; existing estimates of these corrections [52] do not explain systematic differences between precise e^+e^- and τ data. Because of that, most recent analyses of the hadronic vacuum polarization contribution to the muon anomalous magnetic moment rely on the e^+e^- data [1, 17, 29, 30, 31]. Nevertheless, since the τ data played significant role in the recent history of a_μ and, since, it continues to serve as a useful cross-check for the e^+e^- data, it is appropriate to describe how the τ data is incorporated into the analysis of the hadronic vacuum polarization correction to the muon anomalous magnetic moment. We begin with a general discussion of the relation between the hadronic τ decays and the e^+e^- annihilation and then focus on τ decays into two pions since the role of the τ data in reducing the theoretical uncertainty in a_μ^{hvp} is most prominent in the two-pion channel.

Consider the decay of the τ lepton into hadrons. We limit ourselves to processes where the total strangeness S of the hadronic state is zero. The $\Delta S = 0$ interaction of τ leptons with hadrons is described by the Lagrangian

$$\mathcal{L}^{\Delta S=0} = -\frac{V_{ud}^* G_F}{\sqrt{2}}\, \bar{\nu}_\tau \gamma^\mu (1 - \gamma_5)\tau \; \bar{d}\gamma_\mu (1 - \gamma_5)u + h.c. \, , \tag{3.68}$$

where G_F is the Fermi constant and V_{ud} is the element of the Cabbibo–Kobayashi–Maskawa mixing matrix. The amplitude for the decay $\tau^- \to \nu_\tau + X_h$ is given by the matrix element

$$\mathcal{M} = -\frac{iV_{ud}^* G_F}{\sqrt{2}}\, \bar{u}_\nu \gamma^\mu (1 - \gamma_5)u_\tau \; \langle X_h | J_\mu^{\mathrm{hadr}} |0\rangle \, , \tag{3.69}$$

where

$$J_\mu^{\mathrm{hadr}} = V_\mu^- - A_\mu^- \tag{3.70}$$

is the charged hadronic current. Following Sect. 3.2, we use $V_\mu^- = \bar{d}\gamma_\mu u$ and $A_\mu^- = \bar{d}\gamma_\mu \gamma_5 u$ to denote its vector and axial components.

To compute the differential decay rate $\mathrm{d}\Gamma_\tau/\mathrm{d}s$, where $\sqrt{s}$ is the hadronic invariant mass, we square the matrix element in (3.69) and integrate over all final states with the constraint $p_{X_h}^2 = s$. The result can be written as

$$\frac{\mathrm{d}\Gamma_{\tau^-}}{\mathrm{d}s} = \frac{G_F^2 m_\tau^3 |V_{ud}|^2}{16\pi^2}\left(1 - \frac{s}{m_\tau^2}\right)^2 \left[\left(1 + \frac{2s}{m_\tau^2}\right)\mathrm{Im}\Pi_{\mathrm{T}}(s) + \mathrm{Im}\Pi_{\mathrm{L}}(s)\right] \, , \tag{3.71}$$

where

$$\Pi_{\mathrm{T,L}}(s) = \Pi_{\mathrm{T,L}}^V(s) + \Pi_{\mathrm{T,L}}^A(s) \, , \tag{3.72}$$

and the functions $\Pi_{\mathrm{T,L}}^{V,A}(s)$ are defined through correlators of vector and axial currents

$$i\int \mathrm{d}^4 x\, e^{iqx} \langle 0 | T\left\{\mathcal{I}_\mu^-(x)\mathcal{I}_\nu^+(0)\right\} |0\rangle = (q_\mu q_\nu - q^2 g_{\mu\nu})\Pi_{\mathrm{T}}^{\mathcal{I}}(q^2) + q_\mu q_\nu \Pi_{\mathrm{L}}^{\mathcal{I}}(q^2) \, , \tag{3.73}$$

with $\mathcal{I} = V, A$. The imaginary parts of the correlators $\mathrm{Im}\Pi_{\mathrm{T,L}}^{V,A}$ are proportional to the corresponding spectral functions $v_{\mathrm{T,L}}^{V,A}(s)$,

$$v_{\mathrm{T,L}}^{\mathcal{I}}(s) = 2\pi\, \mathrm{Im}\Pi_{\mathrm{T,L}}^{\mathcal{I}}(s) \, . \tag{3.74}$$

In terms of these functions, the differential decay rate for $\tau \to$ hadrons reads

$$\frac{\mathrm{d}\Gamma_{\tau^-}}{\mathrm{d}s} = \frac{G_F^2 m_\tau^3 |V_{ud}|^2}{32\pi^3}\left(1 - \frac{s}{m_\tau^2}\right)^2 \sum_{\mathcal{I}=A,V} \left[\left(1 + \frac{2s}{m_\tau^2}\right) v_{\mathrm{T}}^{\mathcal{I}}(s) + v_{\mathrm{L}}^{\mathcal{I}}(s)\right] . \tag{3.75}$$

The τ data is useful for computing the hadronic vacuum polarization contribution to the muon anomalous magnetic moment because spectral functions measured in τ decays can be employed to predict the hadronic e^+e^-

annihilation cross section. Such relation appears because of the isospin symmetry.

The isospin symmetry is the symmetry of the Lagrangian of strong interactions under rotations in flavor space of up and down quarks. Specifically, consider a doublet of quark fields

$$\psi = \begin{pmatrix} u \\ d \end{pmatrix} \tag{3.76}$$

that transforms in a standard way under flavor SU(2), $\psi' = U\psi$, $U \in$ SU(2). This transformation leaves the low-energy Lagrangian invariant, provided that masses of up and down quarks are equal and the electromagnetic and weak interactions are switched off.

A global symmetry of the Lagrangian implies the existence of conserved currents. In case of the isospin symmetry, these currents are $V_\mu^{\pm,3} = \bar{\psi} T^{\pm,3} \gamma_\mu \psi$, where $T^\pm = T^1 \pm iT^2$ and $T^i = \tau_i/2$ $(i = 1, 2, 3)$ are 2×2 matrices of SU(2) generators with standard commutation relations $[T^a, T^b] = i\epsilon_{abc} T^c$.

The isospin symmetry leads to a host of relations between matrix elements connected by isospin transformations. We are interested in relating the hadronic τ decays and the $e^+ e^-$ annihilation into hadrons. To this end, note that the current $V_\mu^- = \bar{d} \gamma_\mu u$ is just the vector part of the charged hadronic current that enters the calculation of the hadronic decay rate of the τ lepton. Then, using the fact that the vacuum state is the isospin singlet, it is easy to see that

$$\langle 0 | T\{ V_\mu^+(x) V_\nu^-(0) \} | 0 \rangle = 2 \, \langle 0 | T\{ V_\mu^3(x) V_\nu^3(0) \} | 0 \rangle \; . \tag{3.77}$$

We observe that the right hand side of (3.77) is the isovector $I = 1$ part of the electromagnetic current J_μ^{em} since

$$J_\mu^{\mathrm{em}} = \frac{2}{3} \bar{u} \gamma_\mu u - \frac{1}{3} \bar{d} \gamma_\mu d = \frac{1}{6} \bar{\psi} \gamma_\mu \psi + \bar{\psi} T^3 \gamma_\mu \psi = \frac{1}{6} \bar{\psi} \gamma_\mu \psi + V_\mu^3 \; . \tag{3.78}$$

Hence, we may write

$$\langle 0 | T\{ V_\mu^+(x) V_\nu^-(0) \} | 0 \rangle = 2 \, \langle 0 | T\{ J_\mu^{\mathrm{em},I=1}(x) J_\nu^{\mathrm{em},I=1}(0) \} | 0 \rangle \; . \tag{3.79}$$

We stress that this equation holds in the exact isospin limit when the charged vector current V_μ^- is conserved. The current conservation implies that the longitudinal structure function v_{L}^V vanishes. Equation (3.79) relates the transversal spectral function v_{T}^V to the spectral function of the $I = 1$ part of the electromagnetic current, $J_\mu^{\mathrm{em},I=1}$,

$$v_T^V(s) = v^{0,I=1}(s) \; , \tag{3.80}$$

where $v^{0,I=1}$ is defined as

$$v^{0,I=1}(s) = 4\pi \, \mathrm{Im}\, \Pi^{I=1}(s) \; . \tag{3.81}$$

Finally, we note that, for final states consisting of pions, the contribution of vector and axial currents into the τ decay widths $\mathrm{d}\Gamma_\tau/\mathrm{d}s$ can be separated in a simple way: final states with even number of pions are produced by the vector current while final states with odd number of pions are produced by the axial current. This follows from the so-called G-conjugation, a combination of the charge conjugation and the isospin rotation around the second axis by π radians. Pions are eigenstates of the G-conjugation, $G|\pi^{\pm,0}\rangle = -|\pi^{\pm,0}\rangle$. Also, $G = 1$ for the charged vector current and $G = -1$ for the charged axial current. In case of the electromagnetic current, $G = \pm 1$ for its isovector and isoscalar components, respectively.

Hence, we conclude that in the limit of exact isospin symmetry, the measurement of the differential rate $\mathrm{d}\Gamma(\tau \to \nu_\tau + n\pi)/\mathrm{d}s$, where the number of pions n is even, may be used to predict the e^+e^- annihilation cross section $\sigma(e^+e^- \to n\pi)$.[7] If isospin breaking effects are sufficiently small and if the precision of the τ data is higher than or comparable to the precision of the e^+e^- data, we obtain a useful tool for reducing theoretical uncertainty of the hadronic vacuum polarization contribution to the muon anomalous magnetic moment. This is the essence of the suggestion by Alemany, Davier and Höcker [26].

Those arguments play a particular important role for the two-pion final state since, as we explained in Sect. 3.4, it gives the largest contribution to a_μ^{hvp}. Because of that, we specialize to the decay of the τ lepton into two pions, $\tau^- \to \nu_\tau + \pi^- + \pi^0$, in what follows. As we mentioned earlier, the two-pion final state is produced by the charged vector current.

Consider the matrix element of the vector current V_μ^- between the vacuum and the two-pion state. In the limit of exact isospin symmetry, the current V_μ^- is conserved and the matrix element can be described by a single form factor

$$\langle \pi_{p_3}^-, \pi_{p_4}^0 | V_\mu^- | 0 \rangle = \sqrt{2}\, F^-(s)(p_3 - p_4)_\mu \, , \qquad (3.82)$$

where $s = q^2 = (p_3 + p_4)^2$. The transversality of the matrix element (3.82) follows from $q p_3 - q p_4 = m_{\pi^-}^2 - m_{\pi^0}^2 = 0$, in the limit of the isospin symmetry. As a consequence, in that limit, the pions are produced in a P-wave *only*.

Assuming exact isospin symmetry, $F^-(s)$ satisfies the normalization condition $F^-(0) = 1$. Indeed, since the integral $\int \mathrm{d}^3 x V_0^-(x)$ is the lowering generator of flavor SU(2), we may write

$$\langle \pi_p^- | \int \mathrm{d}^3 x V_0^-(x) | \pi_p^0 \rangle = \langle \pi_p^- | L^3 V_0^- | \pi_p^0 \rangle = \sqrt{2}\, \langle \pi_p^- | \pi_p^- \rangle = 2\sqrt{2} p_0 L^3 \, , \quad (3.83)$$

---

[7] Note that this classification applies to truly multipion final states. When processes with the production of η meson in association with n pions are considered, the G-parity violating decay $\eta \to 3\pi$ leads to the $(n + 3)\pi$ final state with G-parity opposite to G-parity of the $\eta + n\pi$ state. Therefore, final states with η-meson require special consideration.

where L^3 is the three-dimensional normalization volume. Comparing this relation with the crossing-continued version of (3.82) at $q = 0$, we verify that $F^-(0) = 1$.

Using the parametrization of the matrix element (3.82), it is straightforward to compute the contribution of the two-pion state to the spectral function $v_{\mathrm{T}}^V(s)$. We obtain

$$v_{\mathrm{T}}^{V,2\pi}(s) = \frac{\beta_-^3}{12}\,|F^-(s)|^2 \, , \tag{3.84}$$

where

$$\beta_-(s) = \frac{2|\boldsymbol{p}|}{\sqrt{s}} = \sqrt{1 - \frac{2m_{\pi^-}^2 + 2m_{\pi^0}^2}{s} + \frac{(m_{\pi^-}^2 - m_{\pi^0}^2)^2}{s^2}} \, . \tag{3.85}$$

The third power of the momentum $|\boldsymbol{p}|$ in (3.84) reflects the fact that the two-pion state produced by a vector current is a P-wave.

For the electromagnetic current, we introduce the pion form factor $F^0(s)$

$$\langle \pi_{p_3}^- \pi_{p_4}^+ | J_\mu^{\mathrm{em}} | 0 \rangle = -F^0(s)(p_3 - p_4)_\mu \, , \tag{3.86}$$

which is also normalized to unity at $s = 0$ by virtue of its relation to the generator of the electric charge $Q = \int \mathrm{d}^3x J_0^{\mathrm{em}}(x)$. The two-pion contribution to the electromagnetic spectral function is

$$v^{0,2\pi}(s) = \frac{\beta_0^3}{12}\,|F^0(s)|^2 \, , \tag{3.87}$$

where $\beta_0 = \sqrt{1 - 4m_{\pi^-}^2/s}$. As we already mentioned, only isovector $I = 1$ part V_μ^3 of J_μ^{em} contributes to the matrix element (3.86) due to G-parity. As the result, the isospin symmetry leads to the equality of the form factors and the two-pion spectral functions,

$$F^-(s) = F^0(s) \, , \qquad v_{\mathrm{T}}^{V,2\pi}(s) = v^{0,2\pi}(s) \, . \tag{3.88}$$

As we see, the isospin symmetry allows us to use data on hadronic τ decays as an independent source of information on the e^+e^- annihilation into hadrons. However, currently, the claimed precision of *both* the e^+e^- and the τ data is such that a very accurate comparison of these data is possible. To make such comparison quantitative, we need to account for isospin violation effects.

3.6 Hadronic Vacuum Polarization: The τ Data and the Isospin Symmetry Violations

The isospin symmetry of strong interactions is not exact. It is broken by the mass splitting of up and down quarks as well as their electromagnetic interactions. Because of the isospin symmetry violation, masses and decay widths

of hadrons that form isospin multiplets are split, form factors of charged
and electromagnetic currents are different and QED radiative corrections to
isospin-connected hadronic final states in the e^+e^- annihilation and τ decays
are unrelated. All this significantly complicates the use of data on τ decays as
a substitute for the e^+e^- annihilation data; in particular, this concerns the
two-pion final state where a percent precision is required. A priori, a magnitude of the isospin violating effects is at the percent level; hence, developing
the theory of isospin violations for the two-pion final state is important.

Consider the decay $\tau^- \to \nu_\tau + \pi^- + \pi^0$. The differential decay rate is given
by (3.75) with $v_T^V \to v_T^{V,2\pi}$ and $v_L^V, v_{T,L}^A \to 0$. It is conventional to normalize
the differential decay rate to the total decay width $\Gamma_{\tau^- \to \nu_\tau^- \pi^- \pi^0} \equiv \Gamma_{\pi^- \pi^0}$.
We obtain

$$\frac{1}{\Gamma_{\pi^- \pi^0}} \frac{\mathrm{d}\Gamma_{\pi^- \pi^0}}{\mathrm{d}s} = \frac{6|V_{ud}|^2}{m_\tau^2} \frac{\Gamma^{(0)}_{\tau^- \to \nu_\tau^- e\bar{\nu}_e}}{\Gamma_{\pi^- \pi^0}} \left(1 - \frac{s}{m_\tau^2}\right)^2 \left(1 + \frac{2s}{m_\tau^2}\right) v_T^{V,2\pi}(s) ,$$

$$(3.89)$$

where we used

$$\Gamma^{(0)}_{\tau^- \to \nu_\tau^- e\bar{\nu}_e} = \frac{G_F^2 m_\tau^5}{192\pi^3} ,$$

$$(3.90)$$

for the tree-level leptonic decay width of the τ lepton.

If all the isospin violating effects are neglected, we extract the spectral
function $v_T^{V,2\pi}(s)$ from (3.89), use (3.88) to obtain the spectral density of
the electromagnetic current and arrive at the prediction for the e^+e^- annihilation cross section into the two-pion state. This leads to the two-pion
contribution to the hadronic vacuum polarization component of the muon
anomalous magnetic moment

$$a_\mu^{\mathrm{hvp},\tau^-}(2\pi) = \frac{\alpha}{\pi} \int\limits_{4m_\pi^2}^{\infty} \frac{\mathrm{d}s}{s} a_\mu^{(1)}(s) \, v_{\mathrm{T,exp}}^{V,2\pi}(s) ,$$

$$(3.91)$$

where the spectral function $v_T^{V,2\pi}$ follows from the τ data as defined by (3.89).
We have put an extra subscript on $v_T^{V,2\pi}$ in (3.91) to emphasize that this
spectral function represents data from τ decays, that are *not corrected for
isospin violations*. When the isospin violating effects are taken into account,
(3.91) is modified; our goal in this section is to derive the correction that has
to be applied to (3.91).

To this end, we have to consider the possibility that isospin symmetry
violations modify the relation between the differential decay rate $\mathrm{d}\Gamma_{\pi^- \pi^0}$ and
the spectral density $v_T^{V,2\pi}$. To find those modifications, we need to consider
the difference of up and down quark masses and the QED corrections. We
note that the QED corrections can be both explicit and implicit – for example,
the charged-neutral pion mass difference is of the electromagnetic origin but
we do not need to derive the splitting and may use the masses of $\pi^\pm$ and π^0
as the input parameters.

Consider the impact of the inequality of up- and down-quark masses on the relation between the τ decay rate and the e^+e^- annihilation cross section. In the derivation of $\Gamma(\tau^- \to \bar{\nu}_\tau^- + \pi^- + \pi^0)$ described earlier, the equality of quark masses was used to argue that the charged vector current is conserved. When quark masses are not equal, this assumption becomes invalid. Nevertheless, it turns out that the matrix element $\langle \pi^- \pi^0 | J_\mu^{\mathrm{hadr}} | 0 \rangle$ is transversal through *first order* in $m_u - m_d$. To see that, consider the divergence of J_μ^{hadr}. We find

$$q^\mu \langle \pi_{p_3}^- \pi_{p_4}^0 \mid J_\mu^{\mathrm{hadr}} \mid 0 \rangle = -i \langle \pi_{p_3}^- \pi_{p_4}^0 \mid \partial^\mu J_\mu^{\mathrm{hadr}} \mid 0 \rangle$$
$$= (m_d - m_u)\langle \pi_{p_3}^- \pi_{p_4}^0 \mid \bar{d}u \mid 0 \rangle = 0 \ , \qquad (3.92)$$

where we used the symmetry under G-conjugation to show that $\langle \pi^- \pi^0 | \bar{d}u | 0 \rangle = 0$. Therefore, even when quark masses are not equal, the longitudinal spectral function v_{L}^V scales as $v_{\mathrm{L}}^V \propto (m_d - m_u)^4$ and can be neglected. In addition, the longitudinal spectral function can also appear due to QED corrections, for example because of the $\pi^{\pm,0}$ mass difference. These corrections affect v_{L}^V only in second order in α and can be safely neglected.

A similar consideration shows that corrections to the transversal spectral function due to unequal quark masses are small, $\Delta v_{\mathrm{T}}^V \propto (m_d - m_u)^2$. Indeed, since the G-parity of the perturbation $(m_d - m_u)(\bar{d}d - \bar{u}u)$ is -1, it can only contribute in second order. The effect *linear* in $m_u - m_d$ appears, however, in the $e^+e^- \to \pi^+\pi^-$ annihilation where the $I = 0$ part of the electromagnetic current starts to contribute. This isospin breaking effect shows up via $\rho - \omega$ mixing which is enhanced by the factor m_ρ/Γ_ρ, thanks to the proximity of ρ and ω masses.

We now turn to explicit QED corrections. Some of them are obvious. For example, (3.89) contains the tree-level leptonic decay width of the τ lepton; it is however conventional to use the experimentally measured value for the leptonic decay width when evaluating (3.89). Then, we should rewrite $\Gamma_{\tau \to \nu_\tau e \bar{\nu}_e}^{(0)}$ through $\Gamma_{\tau \to \nu_\tau e \bar{\nu}_e}^{\mathrm{exp}}$, taking into account the QED radiative corrections. Those corrections are known from earlier computations of the muon life time [53]; we find

$$\Gamma_{\tau^- \to \nu_\tau^- e \bar{\nu}_e}^{\mathrm{exp}} = \Gamma_{\tau \to \nu_\tau^- e \bar{\nu}_e}^{(0)} \left(1 + \delta_{\mathrm{lep}}\right) , \qquad \delta_{\mathrm{lep}} = \frac{\alpha}{2\pi}\left(\frac{25}{4} - \pi^2\right) . \qquad (3.93)$$

To account for this effect one needs to multiply the structure function extracted from τ data using (3.89) by the factor $1 + \delta_{\mathrm{lep}}$,

$$v^{\mathrm{T}}(s) \sim \frac{1}{\Gamma_{\tau \to \nu_\tau e \bar{\nu}_e}^{(0)} \Gamma_{\pi^-\pi^0}} \frac{d\Gamma_{\pi^-\pi^0}}{ds} \sim \frac{1 + \delta_{\mathrm{lep}}}{\Gamma_{\tau \to \nu_\tau e \bar{\nu}_e}^{\mathrm{exp}} \Gamma_{\pi^-\pi^0}} \frac{d\Gamma_{\pi^-\pi^0}}{ds} . \qquad (3.94)$$

Another obvious source of corrections is the mass splitting of charged and neutral pions [54]. These masses enter the factors $\beta_{-,0}^3$ in (3.84, 3.87), and by multiplying the τ spectral function with β_0^3/β_-^3, we account for that effect.

The remaining QED corrections are more involved. To evaluate those corrections, we separate them into virtual and real emission components. The real emission corrections describe the process with an additional photon in the final state, $\tau^- \to \nu_\tau^- + \pi^- + \pi^0 + \gamma$. Since such processes can be distinguished experimentally, if the energy of the emitted photon is not too small, the QED radiative corrections depend on the experimental set-up. In case of τ decays into two pions, the experimental procedure for both ALEPH and CLEO is inclusive with respect to photons; therefore, we have to include the radiative τ decays without restrictions on energies and angles of the emitted photon, when computing QED corrections to the two-pion final state.

Consider first the virtual QED correction to $\tau^- \to \nu_\tau^- + \pi^- + \pi^0$ decay. The τ lepton mass divides the virtuality of the photon k^2 into two regions, $k^2 > m_\tau^2$ and $k^2 < m_\tau^2$. When the photon virtuality is large, the QED corrections are determined by the short-distance properties of the theory since photons with large virtualities resolve the quark structure of the pion. This feature permits a model-independent computation of the QED corrections. The opposite case of small photon virtuality is more complex; we have to deal with pions and other hadrons as degrees of freedom.

Photons with large virtualities renormalize the four-quark operator in the effective weak Lagrangian (3.68). We will show below that their effects are enhanced by the logarithm of the ratio of the W-boson mass to the τ mass and, hence, provide enhanced isospin violating corrections [55].

To compute the renormalization of the effective Lagrangian (3.68) by highly virtual photons, we choose the Landau gauge; the photon propagator reads

$$D_{\alpha\beta}(k) = -\frac{i}{k^2}\,\mathcal{P}_{\alpha\beta}(k)\,, \qquad \mathcal{P}_{\alpha\beta} = g_{\alpha\beta} - \frac{k_\alpha k_\beta}{k^2}\,. \tag{3.95}$$

In this gauge, the vertex corrections as well as the muon and quark wave function renormalization constants are ultra-violet finite; therefore, the only potential source of large logarithms are the two box diagrams where the photon connects the τ lepton with either up or down quark lines, see Fig. 3.6.

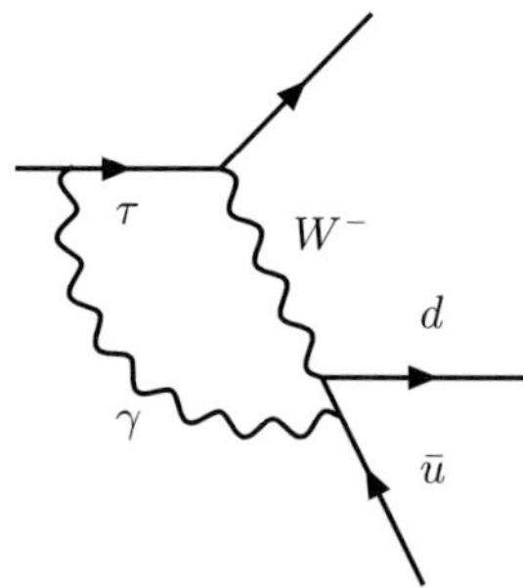

Fig. 3.6. The box diagram that leads to the QED correction to τ decay rate, enhanced by $\ln(m_W/m_\tau)$

Thanks to the W-boson propagator, the integral over Euclidean loop momentum k converges at $k > m_W$; however, the logarithmic sensitivity to m_W remains. To compute the coefficient of the logarithm of the W mass, it is sufficient to assume that the loop momentum k is in the range $\mu_1 < k < \mu_2$, where $m_\tau \ll \mu_1 \ll \mu_2 \ll m_W$. For the loop momentum in that interval, we may neglect all external momenta in Feynman diagrams that describe virtual corrections to the hadronic decay rate of the τ lepton and consider weak interactions as local, approximating the propagator of the W-boson by $ig_{\mu\nu}/m_W^2$.

Among the two box diagrams, only the diagram Fig. 3.6 where the virtual photon is exchanged between the τ lepton and the up anti-quark line contributes; the other box diagram does not produce logarithmically enhanced correction. To see that, we apply the Fierz transformation to the Lagrangian (3.68)

$$\bar{\nu}_\tau \gamma^\mu (1 - \gamma_5)\tau \otimes \bar{d}\gamma_\mu(1 - \gamma_5)u = \bar{\nu}_\tau \gamma^\mu(1 - \gamma_5)u \otimes \bar{d}\gamma_\mu(1 - \gamma_5)\tau \ , \qquad (3.96)$$

and transform the Lorentz structure of the $\tau^- - d$ box diagram into that of a vertex. Since vertex corrections in the Landau gauge are ultra-violet finite, the $\tau^- - d$ box diagram does not produce logarithmically enhanced corrections.

The $\tau^- - \bar{u}$ box diagram, Fig. 3.6, is easy to compute. Within the approximation described above, the expression reads

$$\mathcal{M} = \frac{\alpha G_F V_{ud}^* Q_u Q_\tau}{\sqrt{2}\pi} \int \frac{\mathrm{d}^4 k}{4\pi^2} \frac{\mathcal{P}_{\alpha\beta}(k)}{k^2} \bar{\nu}_\tau \gamma^\mu (1 - \gamma_5) \frac{\hat{k}}{k^2} \gamma^\alpha \tau \ \bar{d}\gamma_\mu(1 - \gamma_5) \frac{\hat{k}}{k^2} \gamma^\beta u \ .$$

$$(3.97)$$

Averaging over directions of the loop momentum and using the identity for the Dirac matrices

$$\gamma^\mu(1 - \gamma_5)\gamma^\alpha\gamma^\beta \otimes \gamma_\mu(1 - \gamma_5)\gamma_\alpha\gamma_\beta = 16\gamma^\mu(1 - \gamma_5) \otimes \gamma_\mu(1 - \gamma_5) \ , \qquad (3.98)$$

we derive

$$\mathcal{M} = -\frac{3 Q_u Q_\tau \alpha}{2\pi} \int_{\mu_1}^{\mu_2} \frac{\mathrm{d}k}{k} \mathcal{M}_0 = -\frac{3 Q_u Q_\tau \alpha}{2\pi} \ln\frac{\mu_2}{\mu_1} \mathcal{M}_0 \ , \qquad (3.99)$$

where $\mathcal{M}_0 = -iV_{ud}^* G_F/\sqrt{2}\ \bar{\nu}_\tau\gamma^\mu(1 - \gamma_5)\tau\bar{d}\gamma_\mu(1 - \gamma_5)u$ is the tree-level amplitude for the semileptonic decay $\tau^- \to \nu_\tau \bar{u}d$.

From (3.99) we see that the QED corrections due to photons with large virtualities $\mu_1 \ll k \ll \mu_2$ do not depend on the kinematics of the process and, effectively, change the coupling constant of the four-fermion operator in the Lagrangian (3.68). It is clear that, with the logarithmic accuracy, we may take $\mu_1 = m_\tau$, $\mu_2 = m_W$. The result is a multiplication of the four-fermion operator in (3.68) by the *short-distance* QED Wilson coefficient [55]

$$S_{\mathrm{EW}} = 1 + \frac{\alpha}{\pi} \ln \frac{m_W}{m_\tau} \ . \tag{3.100}$$

As a consequence, the hadronic decay rate of a τ lepton is increased by S_{EW}^2. Numerically,

$$S_{\mathrm{EW}}^2 \approx 1.018 \ , \tag{3.101}$$

which implies that the short-distance QED effects increase the τ decay rate into *any* hadronic channel by, approximately, two percent. Apart from the limitations of the logarithmic accuracy, (3.101) is one of a few model-independent results related to QED corrections to hadronic decays of the τ-lepton.

We point out that the QED corrections to the *total* hadronic decay width of a τ-lepton can be reliably calculated *beyond* the logarithmic accuracy. Such possibility occurs because in the inclusive width the long-distance effects in emission of real photons cancel against long-distance effects in the virtual corrections, so that the QED corrections to the total hadronic decay rate are determined by photons with large, $\sim m_\tau^2$, virtualities. This cancelation of infra-red logarithms in the inclusive width follows from Kinoshita–Lee–Nauenberg theorem [56, 57]. The QED corrections to the total decay rate were computed in [58]; they read

$$\Gamma_{\tau^- \to \nu_\tau + \mathrm{hadr}} = \Gamma_{\tau^- \to \nu_\tau + \mathrm{hadr}}^{(0)} \left[1 + \frac{\alpha}{\pi} \left(\ln \frac{m_Z^2}{m_\tau^2} + \frac{85}{24} - \frac{\pi^2}{2} \right) \right] . \tag{3.102}$$

Although long-distance effects cancel in the total decay width, this does not necessarily happen in less inclusive observables. For example, the hadronic invariant mass distribution, (3.89), is affected by the long-distance effects. The weights with which this distribution is integrated are different for the inclusive widths and for the hadronic corrections to the muon magnetic anomaly; in the latter case, the small mass region is emphasized so that the cancelation of long-distance effects is, in general, destroyed. For this reason it is not possible to compute the QED corrections to the hadronic invariant mass distribution in a *fully* model-independent way since hadronic dynamics has to be accounted for. However, it is possible to understand essential features of the QED corrections in a way that is model-independent, to a large extent. The most comprehensive analysis of the QED effects in τ decays is given in [52]. While we do not follow the discussion in that reference, we comment on essential features of that analysis below.

Before diving into that discussion, we note that a theoretical limit exists when the cancelation of long-distance effects in QED corrections occurs in a_μ^{hvp} so that the QED corrections that have to be applied to τ data to derive the $e^+ e^-$ annihilation cross section can be computed exactly. This situation occurs if the total decay rate of the τ lepton is dominated by the two-pion channel which, in turn, is dominated by the ρ meson contribution. We further assume that the ρ meson is *narrow*. In such a limit, hadronic invariant mass distribution becomes a delta-function, $\delta(s - M_\rho^2)$, even when the QED corrections are accounted for. The coefficient of this delta function is *uniquely* determined from the QED corrections to the total decay rate of the τ-lepton, (3.102). It is then obvious that the long-distance effects cancel in the calculation of the muon magnetic anomaly in the same way as they do in the calculation of the inclusive width. This argument implies *small* QED corrections since the possible enhancement factors associated with long-distance physics do not play a role under such circumstances. In a realistic case of a finite ρ meson width, the situation is, clearly, more complex. We discuss below the long-distance enhancement in the hadronic invariant mass spectrum in hadronic τ decays and the degree of cancelation of long-distance effect in a calculation of a_μ^{hvp} in models where the width of the ρ meson has its actual value.

An enhancement of long-distance effects in QED radiative corrections to $\mathrm{d}\Gamma_{\pi^-\pi^0}/\mathrm{d}s$, where s is the invariant mass of the charged and neutral pions, is seen from the fact that these corrections are logarithmically sensitive to the mass of the pion; interestingly, they can be derived in a model-independent way. This happens because the terms enhanced by $\ln m_\pi$ originate entirely from collinear photon emissions off the external pion leg; since the emission is collinear, the pion remains very close to the mass-shell and there is no influence of the pion structure on the radiation pattern.

It is easy to write down a formula for the QED radiative corrections in the logarithmic approximation. Define $z = (s - 4m_\pi^2)/(m_\tau^2 - 4m_\pi^2)$. Then, the QED corrections read

$$\frac{\mathrm{d}\Gamma^{(1)}_{\pi^-\pi^0}}{\mathrm{d}z} = \frac{\alpha}{\pi} \ln \frac{\langle E_\pi \rangle}{m_\pi} \int_z^1 \frac{\mathrm{d}\xi}{\xi} \, P_{\pi\pi}\left(z/\xi\right) \frac{\mathrm{d}\Gamma^{(0)}_{\pi^-\pi^0}}{\mathrm{d}\xi} \, , \tag{3.103}$$

where the average energy of the charged pion in the decay $\tau \to \nu_\tau + \pi^- + \pi^0$ is estimated to be

$$\langle E_\pi \rangle = \frac{m_\tau^2 + s}{4m_\tau} \, . \tag{3.104}$$

The function $P_{\pi\pi}$, given by

$$P_{\pi\pi}(z) = \left[\frac{2}{1-z}\right]_+ - 2 + 2\delta(1-z) \, , \tag{3.105}$$

describes the collinear splitting $\pi \to \pi + \gamma$. The plus-distribution $[1/(1-z)]_+$ is defined as

$$\int_0^1 \mathrm{d}z f(z) \left[\frac{1}{1-z}\right]_+ = \int_0^1 \mathrm{d}z \, \frac{f(z) - f(1)}{1-z} \, , \tag{3.106}$$

for an arbitrary smooth function $f(z)$. Thanks to that relation, the logarithmically enhanced terms in (3.103) disappear once the integral over z is performed,

$$\int_0^1 \mathrm{d}z \; P_{\pi\pi}(z) = 0 \, . \tag{3.107}$$

As a consequence, there are no $\ln(m_\pi)$-enhanced corrections to the total decay rate of the τ lepton, as required by the Kinoshita-Lee-Nauenberg theorem [56, 57]. On the other hand, (3.103) shows that the QED corrections can significantly distort the invariant mass spectrum of the two pions.

Since the $\ln(m_\pi)$ enhancement factor is not large, the non-logarithmic corrections neglected in (3.103) may be important numerically. Nevertheless, it is useful to isolate $\ln(m_\tau/m_\pi)$-enhanced QED corrections to the two-pion decay rate of the τ lepton because i) they are model-independent and ii) the presence of $\ln(m_\pi)$-enhanced terms is the feature that distinguishes QED radiative corrections to the two-pion final state in τ decays and in e^+e^- annihilation. The expression for the QED radiative corrections (3.103) allows us to understand qualitative features of these corrections.

To quantify the impact of the QED corrections on the differential decay rate, we define the function $G_{\mathrm{EM}}(s)$ through

$$\frac{\mathrm{d}\Gamma_{\pi^-\pi^0}}{\mathrm{d}s} = \frac{\mathrm{d}\Gamma^{(0)}_{\pi^-\pi^0}}{\mathrm{d}s} \, G_{\mathrm{EM}}(s) \, , \tag{3.108}$$

where $\mathrm{d}\Gamma^{(0)}$ is the differential decay rate without QED corrections, (3.89). The function $G_{\mathrm{EM}}(s)$ depends on the parametrization of the pion form factor; we use the Gounaris–Sakurai parametrization obtained from a combined fit to the e^+e^- and τ data [59]. The explicit form of the form factor as well as the input parameters are given in Appendix A.2.

We plot the function $G_{\mathrm{EM}}(s)$ in Fig. 3.7. The behavior of the QED corrections is typical to a situation when the tree-level matrix element describes the production of a narrow resonance; the correction, negative for $s \approx m_\rho^2 \approx 0.6 \text{ GeV}^2$, rapidly becomes positive for $\sqrt{s}$ smaller than the mass of the ρ-meson. Away from the resonance peak, the corrections can be as large as a few percent.

Our result for $G_{\mathrm{EM}}(s)$ reproduces fairly well the complete calculation of the QED radiative corrections to the two-pion decay rate of the τ lepton reported in [52]; the agreement gets unsatisfactory for small values of s where the $\mathcal{O}(\ln m_\pi)$ correction overestimates the full result. Since, as we mentioned already, it is not expected that keeping only terms enhanced by $\ln m_\pi$ gives an accurate description of the spectrum, we use $G_{\mathrm{EM}}(s)$ derived in [52] for evaluating the impact of the QED corrections on the charged pion form factor. Nevertheless, the proximity of the two calculations ensures that the shape of the two-pion invariant mass distribution in τ decays is governed by the collinear QED radiation that does not depend on the pion structure and, therefore, can be understood within the framework of scalar QED. The fact that the structure-dependent radiation does not significantly contribute

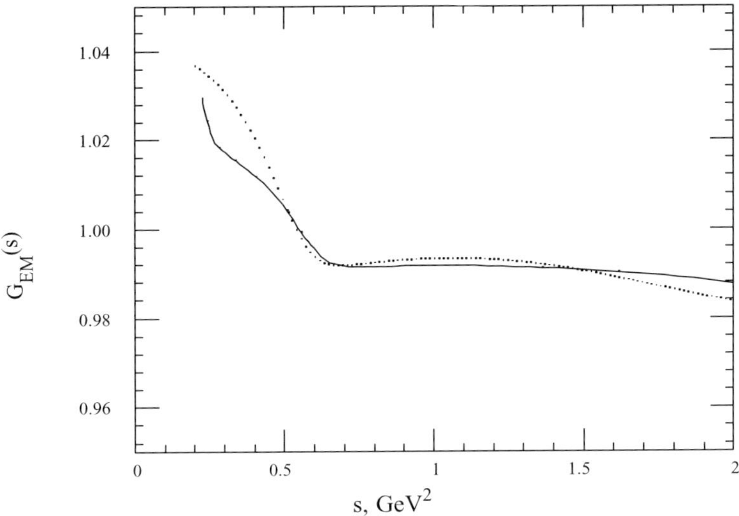

Fig. 3.7. The function $G_{\mathrm{EM}}(s)$. The solid line is the complete calculation [52], the dashed line is the approximation (3.103)

to the QED radiative corrections to τ decays into two pions was explicitly demonstrated in [52].

The results for the QED corrections to semileptonic τ decays into two pions allow us to derive the form factor $F^-(s)$ from the differential decay rate. Using (3.89, 3.108), we find

$$|F^-(s)|^2 = |F^-_{\exp}(s)|^2 \frac{1 + \delta_{\mathrm{lep}}}{S^2_{\mathrm{EW}} G_{\mathrm{EM}}(s)} \ . \tag{3.109}$$

As we already mentioned, this formula is slightly misleading, since the electromagnetic correction factor $G_{\mathrm{EM}}(s)$ depends on $F^-(s)$. However, because this dependence is weak, we may ignore it to the first approximation.

Equation (3.109) gives the form factor of the charged hadronic current with the isospin I equal to one. To compute the e^+e^- annihilation cross section, we need the pion form factor that contains both isovector and isoscalar components. The major contribution to the $I = 0$ component is the isospin violating mixing of ρ and ω mesons; it can be accounted for by using parameters of the ω resonance determined from the e^+e^- annihilation. In addition, there are *internal* isospin violating effects in $I = 1$ form factor related, for example, to differences in masses and widths of the charged and neutral ρ-mesons. To indicate the presence of such effects, we introduce a function $R_f(s)$ and write

$$|F^0(s)|^2 = |F^-(s)|^2 \left|1 + \delta \frac{s}{M^2_\omega} BW_\omega(s)\right|^2 R_f(s) \ , \tag{3.110}$$

where $BW_\omega(s)$ is the Breit-Wigner parametrization of the ω resonance (see Appendix A.2).

We now use (3.87) that defines the neutral spectral density v_0 and (3.109, 3.110, 3.89) to derive the relation between the "experimental" form factor $F^-_{\exp}(s)$ and the e^+e^- annihilation cross section into two pions. Note that the cross section derived in this way is the *bare* cross section, with all the QED corrections to $\gamma^* \to \pi^+\pi^-$ excluded. However, as we explained at the beginning of this chapter, this is not sufficient; the QED corrections to $e^+e^- \to \pi^+\pi^-$, including processes with an additional photon in the final state, have to be included for the two-pion channel. We parametrize those corrections by a function $\Lambda_{fs}(s)$ and write

$$\sigma_{\pi^+\pi^-} = \frac{\pi\alpha^2}{3s}\,\beta_0^3\,|F^0(s)|^2\left(1 + \frac{\alpha}{\pi}\Lambda_{fs}(s)\right). \tag{3.111}$$

Using this equation together with (3.110), we derive the result for the two-pion contribution to a_μ^{hvp}

$$a_\mu^{\mathrm{hvp}}(2\pi) = \frac{\alpha}{12\pi}\int\limits_{4m_\pi^2}^{\infty}\frac{\mathrm{d}s}{s}\,a_\mu^{(1)}(s)\beta_-^3(s)|F^-_{\exp}(s)|^2 C(s)\,, \tag{3.112}$$

where the function $C(s)$ reads

$$C(s) = \frac{\beta_0^3}{\beta_-^3}\,\frac{1 + \delta_{\mathrm{lep}}}{S_{\mathrm{EW}}^2 G_{\mathrm{EM}}(s)}\left|1 + \delta\,\frac{s}{M_\omega^2}\,BW_\omega(s)\right|^2 R_f(s)\left(1 + \frac{\alpha}{\pi}\Lambda_{fs}(s)\right). \tag{3.113}$$

Comparing (3.91, 3.112), we find the correction that has to be applied to the evaluation of the two-pion contribution to the hadronic vacuum polarization component of a_μ if the τ data uncorrected for the isospin violating effects is used in the original derivation

$$\delta a_\mu^{\mathrm{hvp},\tau}(2\pi) = \frac{\alpha}{12\pi}\int\limits_{4m_\pi^2}^{\infty}\frac{\mathrm{d}s}{s}\,a_\mu^{(1)}(s)\,\beta_-^3(s)|F^-_{\exp}(s)|^2\,(C(s) - 1)\,. \tag{3.114}$$

Before discussing numerical implications of the isospin violation correction (3.114), some clarifications are in order. As we already mentioned, for numerical evaluations we use Gounaris–Sakurai parametrization of the pion form factor. The explicit parametrization of the form factor as well as numerical values for the masses of vector resonances, their widths etc. are given in Appendix A.2. The function $\Lambda_{fs}(s)$ is known in scalar QED [60]

$$
\begin{aligned}
\Lambda_{fs}(s) \;=\; & \frac{1+\beta^2}{\beta}\left\{4\mathrm{Li}_2\left(\frac{1-\beta}{1+\beta}\right) + 2\mathrm{Li}_2\left(-\frac{1-\beta}{1+\beta}\right)\right. \\
& \left. - 3\ln\left(\frac{2}{1+\beta}\right)\ln\frac{1+\beta}{1-\beta} - 2\ln\beta\ln\frac{1+\beta}{1-\beta}\right\} - 3\ln\frac{4}{1-\beta^2} - 4\ln\beta \\
& + \frac{1}{\beta^3}\left(\frac{5}{4}(1+\beta^2)^2 - 2\right)\ln\frac{1+\beta}{1-\beta} + \frac{3}{2}\frac{1+\beta^2}{\beta^2}\,,
\end{aligned}
\tag{3.115}
$$

where $\beta = \beta_0(s)$. It turns out that the QED radiative corrections described by $\Lambda_{fs}(s)$ are large; to see this, note that setting $m_\pi = 0$ leads to $\Lambda_{fs}(s) = 3$. Since the major contribution to a_μ^{hvp} comes from $s \approx M_\rho^2 \gg m_\pi^2$, it is unlikely that scalar QED accurately describes radiative corrections at such virtualities. A possible way to check the sensitivity of $\Lambda_{fs}(s)$ to the pion structure is to compute it using the vector meson dominance model where the coupling of photons to pions is modified by introducing a form factor. Such calculation, for zero pion mass, was reported in [61]; the result is a small suppression of the asymptotic value of the function $\Lambda_{fs}(s) = 3 \to 9/4$. Note, however, that the validity of the above estimates can be questioned if we assume that the two-pion contribution to a_μ is saturated by the narrow ρ-meson. In that case, the $e^+e^- \to \pi^+\pi^-$ contribution to a_μ is determined by an integral of the $e^+e^- \to \pi^+\pi^-$ annihilation cross section over sufficiently large energy range. Because of the quark-hadron duality, the latter quantity can be calculated by integrating the $e^+e^- \to$ quarks annihilation cross-section. In that case, the QED correction to the annihilation cross-section is computed with quarks. Keeping only the isovector part of the electromagntic current, we derive

$$
\Lambda_{fs} = \frac{3}{4}\sum_{q=u,d} Q_q^2 = \frac{5}{12} \approx 0.5\,,
\tag{3.116}
$$

which is much smaller than $\Lambda_{fs} \approx 3$ obtained in the point-pion approximation. Keeping in mind that the main contribution to a_μ indeed comes from $\sqrt{s} \approx m_\rho$, the mismatch of the two estimates implies breaking of quark-hadron duality at (marginally) short distances.

Given the differences in Λ_{fs} computed with point-like pions and quarks, we have to decide which of the two estimates is more appropriate; we argue that it is a combination of the two. A good way to estimate the correction that needs to be applied to the τ data due to the final state interaction in $e^+e^- \to \pi^+\pi^-$ is the following. We may use (3.115) to describe the final state interaction of the $\pi^+\pi^-$ pair in the two-pion threshold region, $s \leq m_\rho^2/2$ and use (3.116) in the region $s > m_\rho^2/2$. This leads to the contribution to $\delta a_\mu^{\mathrm{hvp},\tau}$ of about 20×10^{-11}; we use this result in Table 3.4 below.

Finally, we need to specify the function $R_f(s)$ that is introduced to account for isospin violating effects in the $I = 1$ part of the pion form factor. A possible approach is to compute the form factors $F^{-,0}(s)$ in unitarized chiral perturbation theory [52, 62]; then, the isospin violating effects in the form

Table 3.4. Various contributions to $\delta a_\mu^{\mathrm{hvp}}$

Source of isospin violation	$\delta a_\mu^{\mathrm{hvp},\tau^-} \times 10^{11}$
$(1 + \delta_{\mathrm{lep}})\, S_{\mathrm{EW}}^{-2}$	-119
G_{EM}^{-1}	-10
β_0^3/β_-^3	-90
$\rho - \omega$ mixing	24
final state $\mathcal{O}(\alpha)$	20
$\Gamma_{\rho^-} \neq \Gamma_{\rho^0}$	5
$m_{\pi^-} \neq m_{\pi^0}$ in $F(s)$	-8
$m_{\rho^0} \neq m_{\rho^-}$	7

factors appear explicitly and are given in terms of the mass and width differences of charged and neutral ρ mesons and some low-energy constants of chiral perturbation theory [52]. Unfortunately, those effects can not be estimated theoretically in a reliable way. Indeed, the mass difference of charged and neutral ρ is estimated theoretically [63] to be $-0.7\,\mathrm{MeV} < m_{\rho^+} - m_{\rho^0} < 0.4\,\mathrm{MeV}$, but, as pointed out in [64], this is not supported by the latest experimental data. For example, Davier [59] finds from a combined fit to the e^+e^- and the τ data that $m_{\rho^+} - m_{\rho^0} = (2.3 \pm 0.8)\,\mathrm{MeV}$. In addition, with the parameters used in [52], the theoretical difference between the widths of charged and neutral ρ-mesons is about 2 MeV, whereas experimental data seems to point out to a negligible width difference.

Hence, it appears difficult to give an accurate *theoretical* description of the isospin violating effects in the pion form factor. We therefore model those effects by assuming Gounaris–Sakurai parametrization of the form factor and using masses and widths for charged and neutral ρ-mesons, obtained from fits to the e^+e^- and the τ data. Of course, doing so defies the goal of computing $a_\mu^{\mathrm{hvp},\tau}$ *without* the e^+e^- data, since we need these data to estimate $R_f(s)$. However, it is instructive to do so, to get an idea about the magnitude of the effect.

We turn to numerical estimates of the isospin violating corrections. To exhibit the relative importance of various isospin violation corrections, it is convenient to display different contributions to $\delta a_\mu^{\mathrm{hvp},\tau}(2\pi)$ separately; this is done in Table 3.4. As we remarked earlier, for $F_{\mathrm{exp}}^-(s)$, we use Gounaris–Sakurai parametrization; the explicit form of the form factor and numerical values of masses and widths are given in Appendix A.2. Adding the entries in Table 3.4, we derive $\delta a_\mu^{\mathrm{hvp},\tau}(2\pi) = -170 \times 10^{-11}$ which is the shift that one has to apply if the τ data is used to evaluate the two-pion contribution to the hadronic vacuum polarization correction to the anomalous magnetic moment. Note that this correction is rather close to the correction in the first row in Table 3.4, which is an unambiguous short-distance result. This proximity of the two numbers implies that there are significant cancelations of different *long-distance* effects that affect the relation between the two-pion decay rate of the τ lepton and the $e^+e^- \to \pi^+\pi^-$ annihilation cross section. In particular, it is reflected in the small correction due to G_{EM}^{-1} which, as we argued earlier, can be understood in the limit when the ρ meson width is small.

To estimate the uncertainty in $\delta a_\mu^{\mathrm{hvp},\tau}$ we have to analyze how well different isospin symmetry corrections shown in Table 3.4 are known. While some of them, such as the kinematic correction β_0^3/β_-^3 and the correction caused by the $\rho - \omega$ mixing are known accurately, the others are not. For example, the evaluation of the short distance correction $(1 + \delta_{em})S_{\mathrm{EW}}^{-2}$ is done with the choice of the factorization scale $\mu = m_\tau$ (cf. (3.100)). However, we point out that, in spite of the fact that the scale μ is an arbitrary separation scale and, hence, the dependence on it must cancel, in practice it does not. We can estimate the uncertainty in the short distance correction by evaluating it at $\mu = m_\tau/2$. This choice leads to $(1 + \delta_{em})S_{\mathrm{EW}}^{-2} = -136 \times 10^{-11}$, a shift by -17×10^{-11}. To estimate the uncertainty in the evaluation of the long-distance QED corrections to hadronic invariant mass distribution in τ-decays, we note that if we treat ρ as a narrow resonance and utilize the corrections to the total decay rate (3.102), we find that the $G_{\mathrm{EM}}^{-1}(s)$ contribution to $\delta a_\mu^{\mathrm{hvp},\tau}$ turns out to be $+14 \times 10^{-11}$. On the other hand, if the long-distance QED corrections to hadronic invariant mass spectrum are computed with the function $G_{\mathrm{EM}}(s)$ calculated in the logarithmic approximation, (3.103), the contribution to $\delta a_\mu^{\mathrm{hvp},\tau}$ turns out to be $\sim -30 \times 10^{-11}$. We consider the differences between these estimates and the result -10×10^{-11} in Table 3.4 to be indicative of the accuracy with which the long-distance QED effects in hadronic invariant mass spectrum can be computed; as the result, we take $\pm 20 \times 10^{-11}$ as the uncertainty estimate.

Another source of uncertainty is the final state $\mathcal{O}(\alpha)$ correction. As we discussed earlier, calculations with point-like pions differ from calculations where the pion structure is introduced through vector meson dominance by, approximately, 12×10^{-11}. An even stronger discrepancy occurs when the calculation of Λ_{fs} with point-like pions is compared with a similar calculation that utilizes quarks as degrees of freedom. To estimate the uncertainty on this contribution we assume that the quark-based calculation may be incorrect up to a factor of two and also assign $\pm 5 \times 10^{-11}$ uncertainty to the contribution due to two-pion threshold region to $\mathcal{O}(\alpha)$ correction in e^+e^- annihilation. The uncertainty on the final state radiative correction then becomes 10×10^{-11}. Finally, note that the width difference between Γ_{ρ^-} and Γ_{ρ^0} of 1 MeV leads to a shift of $\delta a_\mu^{\mathrm{hvp},\tau}$ by 30×10^{-11}.

To estimate the total uncertainty in $\delta a_\mu^{\mathrm{hvp},\tau}$, we note that the uncertainties described above are systematic; hence, adding those uncertainties in quadratures may underestimate the total uncertainty while adding them linearly, likely, overestimates it. We choose $\delta a_\mu^{\mathrm{hvp},\tau} = \pm 60 \times 10^{-11}$ as an uncertainty estimate; this value is between the two extreme cases. Hence, we arrive at the result $\delta a_\mu^{\mathrm{hvp},\tau} = -170(60) \times 10^{-11}$.

Our result for $\delta a_\mu^{\mathrm{hvp},\tau}$ differs somewhat from similar results reported in e.g. [52, 17]; for example, in [17] $\delta a_\mu^{\mathrm{hvp},\tau}$ was estimated to be $-93(20) \times 10^{-11}$; the major reason for this difference is the latest fit to the e^+e^- and the τ data [59] that shows that the widths of charge and neutral ρ mesons are, practically, the same.

Up to now, we have been discussing the influence of the τ data on the hadronic vacuum polarization contribution to a_μ. On the other hand, it is possible to compare directly the charged and neutral pion form factors for various values of s, after all the isospin violation effects have been accounted for. The result of such a comparison is shown in Fig. 3.8. It follows that the difference between the e^+e^- and the τ data is not uniform and that large $\sim 10\%$ difference occurs to the right of $s = m_\rho^2 \approx 0.6$ GeV^2. Usually, Fig. 3.8 is presented as the strong evidence that the τ and the e^+e^- data are not compatible even after the isospin violation effects are taken into account.

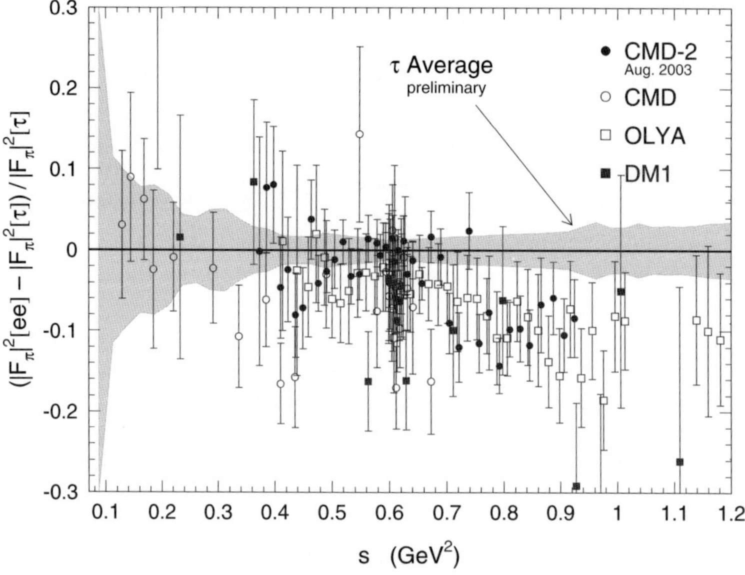

Fig. 3.8. The comparison of the pion form factors obtained from the e^+e^- and the τ data with all the known isospin corrections applied. The band shows the uncertainties in the τ-data, [65]. The significant mass difference of charge and neutral ρ mesons was not considered in this plot

However, there is a very significant difference between computing *integrated* quantities, like a_μ^{hvp} and making the point-by-point comparison of the charged and neutral pion form factor. This is so because, in the latter case, the result depends very strongly on possible mass and width differences of the charged and neutral ρ-mesons [64], whereas in case of a_μ^{hvp}, as follows from Table 3.4, the mass difference of the ρ-mesons leads to a minor effect, only.

Finally, we note that our calculation of $\delta a_\mu^{\mathrm{hvp},\tau}$ essentially removes the tension between the e^+e^-- and τ-based calculations of hadronic vacuum polarization contribution to a_μ^{hvp}. For example, [41] gives $a_\mu^{\mathrm{hvp},\tau}(2\pi) - a_\mu^{\mathrm{hvp},e^+e^-}(2\pi) = (119 \pm 60) \times 10^{-11}$; this result is obtained with $\delta a_\mu^{\mathrm{hvp},\tau} = -93(20) \times 10^{-11}$. With our result for $\delta a_\mu^{\mathrm{hvp},\tau}$, the difference between the two computations of a_μ^{hvp} is, practically, removed

$$a_\mu^{\mathrm{hvp},\tau}(2\pi) - a_\mu^{\mathrm{hvp},e^+e^-}(2\pi) = (40 \pm 50_{\mathrm{exp}} \pm 60_{\mathrm{th}}) \times 10^{-11} \ . \qquad (3.117)$$

The two errors in (3.117) reflect the experimental uncertainty in the data as well as the theoretical uncertainty related to the application of the isospin symmetry violation correction to the τ data.

Before concluding this section, we stress that, in spite of the fact that (3.117) indeed resolves the disagreement between the e^+e^- and the τ data, the τ data still should not be used as independent data for the purpose of a_μ^{hvp} evaluation. The reason is that the isospin violation effects in the pion form factor are significant. Existing theoretical calculations of isospin violating corrections capture large long-distance effects but these effects cancel out when $\delta a_\mu^{\mathrm{hvp},\tau}$ is computed. What remains are the $\mathcal{O}(\alpha/\pi)$ effects that can not be reliably computed from first principles. Although these effects are not enhanced, with the current level of precision they lead to considerable uncertainties when τ data, corrected for isospin violation, is confronted with e^+e^- data. Therefore, even if we gain in experimental precision by using the τ data, we loose in theoretical clarity which is indispensable for making convincing calculation of a_μ^{hvp}.

3.7 Higher Order QED Corrections to Hadronic Vacuum Polarization Contribution to a_μ

An interesting conclusion from the discussion in the previous section is the fact that the QED corrections to a_μ^{hvp} are indeed at the percent level, which is approximately 50×10^{-11}. Since this number is comparable to the experimental uncertainty in the muon anomalous magnetic moment, we have to account for all the QED effects that might affect the hadronic contribution to a_μ. There are two such effects that we have not considered so far; the hadronic light-by-light scattering contribution to a_μ that is studied in Chap. 6 and the two-loop QED diagrams with hadronic vacuum polarization insertion, shown in Fig. 3.9. We discuss those diagrams in this section.

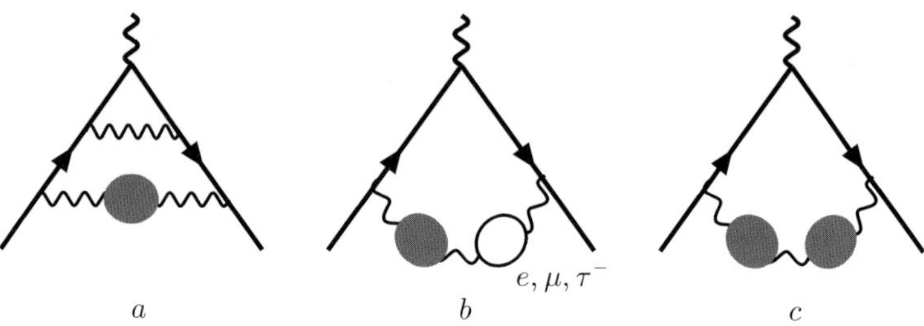

Fig. 3.9. QED corrections to hadronic vacuum polarization contribution to the muon anomalous magnetic moment

There are three classes of diagrams that are usually considered separately; they are shown in Fig. 3.9. The first class, Fig. 3.9a, includes diagrams that have a single hadronic vacuum polarization insertion on one of the photons lines in two-loop QED corrections to the muon magnetic anomaly and that do not have any additional vacuum polarization insertion due to either electron or tau lepton. The diagrams with single electron, muon or tau vacuum polarization and the hadronic vacuum polarization form the second class, Fig. 3.9b. The diagram with two hadronic vacuum polarizations is referred to as the third class. The contribution of those diagrams to the muon anomalous magnetic moment can be computed using representations of the form

$$a_\mu^{\text{hvp,NLO,(i)}} = \frac{\alpha^3}{3\pi} \int\limits_{4m_\pi^2}^{\infty} \frac{ds}{s} R^{\text{hadr}}(s) K^i(s) , \qquad i = a, b ;$$

$$a_\mu^{\text{hvp,NLO,(c)}} = \frac{\alpha^3}{9\pi} \int\limits_{4m_\pi^2}^{\infty} \int\limits_{4m_\pi^2}^{\infty} \frac{ds}{s} \frac{ds'}{s'} R^{\text{hadr}}(s) R^{\text{hadr}}(s') K^c(s, s') .$$

$$(3.118)$$

The analytic results for the kernels in a form convenient for numerical integration can be found in [66]. We quote numerical values from a recent evaluation [1]

$$a_\mu^{\text{hvp,NLO}} = [-207.3(1.8)_a + 106(1)_b + 3.4(1)_c] \times 10^{-11} = -98(1) \times 10^{-11} ,$$

$$(3.119)$$

where the contributions from classes a, b, c are displayed separately. As we see, the NLO QED corrections to hadronic vacuum polarization are close to 1.5%; the contribution of class c, the double hadronic vacuum polarization contribution to a_μ, is very small, as expected.

3.8 Total Hadronic Vacuum Polarization Contribution

In this section we summarize the results of our discussion of the hadronic vacuum polarization contribution. For reasons explained in the main body of this chapter, we will not use the results based on the τ data. We point out, however, that the e^+e^- and the τ based evaluations of a_μ^{hvp} seem to be consistent within the uncertainties related to the experimental data and the lack of theoretical understanding of the isospin symmetry violations.

The results for a_μ^{hvp} are compiled in Table 3.1. It is clear that the most recent evaluations [1, 29, 17, 31, 30] produce results that overlap very well within the assigned error bars. Reference [29] is an update of [17], that includes the recent $e^+e^- \to \pi^+\pi^-$ data from KLOE that shifts the result of [17] by -27×10^{-11} and makes it closer to [1]. As we explained in the text, the difference between [17] and [1] is *not related* to the $e^+e^- \to \pi^+\pi^-$ channel and, hence, the *complete* agreement of [29] and [1] is misleading. For the Standard Model value of the muon anomalous magnetic moment, we employ the leading order hadronic vacuum polarization contribution from [29],

$$a_\mu^{\mathrm{hvp}} = 6934(63) \times 10^{-11} \,. \tag{3.120}$$

However, note that new results on $e^+e^- \to \pi^+\pi^-$ by SND collaboration as well as updated results of CMD-2 collaboration were not included in the analysis of [29]. As follows from the discussion in Sect. 3.4.5, inclusion of these data is likely to increase a_μ^{hvp} in (3.120) by about one standard deviation.

For the NLO QED corrections to the hadronic vacuum polarization, we use [1]

$$a_\mu^{\mathrm{hvp,NLO}} = -98(1) \times 10^{-11} \,. \tag{3.121}$$

The total hadronic vacuum polarization contribution is given by the sum of (3.120, 3.121).

References

1. K. Hagiwara, A. Martin, D. Nomura and T. Teubner, Phys. Rev. D **69**, 093003 (2004).
2. G. 't Hooft, Nucl. Phys. B **72**, 461 (1974); **75**, 461 (1974).
3. S. Coleman, $1/N$, in *Aspects of Symmetry*, Cambridge University Press, Cambridge, 1985;
   A. Manohar, Les Houches lectures on *Large N QCD*, hep-ph/9802419.
4. K. Wilson, Phys. Rev. **179**, 1499 (1969);
   K. Wilson and J. Kogut, Phys. Reports **12**, 75 (1974).
5. M. A. Shifman, A. I. Vainshtein and V. I. Zakharov, Nucl. Phys. B **147**, 385 (1979); **147**, 448 (1979).
6. D. J. Gross and F. Wilczek, Phys. Rev. Lett. **30**, 1343 (1973).
7. H. Politzer, Phys. Rev. Lett. **30**, 1346 (1973).
8. C. G. Callan, Phys. Rev. D **2**, 1541 (1970);
   K. Symanzik, Comm. Math. Phys. **18**, 227 (1970).
9. V. A. Novikov, M. A. Shifman, A. I. Vainshtein and V. I. Zakharov, Fortsch. Phys. **32**, 585 (1984).
10. B. L. Ioffe, Phys. Atom. Nucl. **66**, 30 (2003) [Yad. Fiz. **66**, 32 (2003)].

11. M. Gell-Man, R. J. Oakes and B. Renner, Phys. Rev. **175**, 2195 (1968).
12. T. Das, G. S. Guralnik, V.S. Mathur, F. E. Low and J. E. Young, Phys. Rev. Lett. **18**, 759 (1967).
13. M. Knecht, S. Peres and E. de Rafael, Phys. Lett. B **443**, 255 (1998).
14. R. Dashen, Phys. Rev. **183**, 1245 (1969).
15. S. Weinberg, Phys. Rev. Lett. **18**, 507 (1967).
16. E. de Rafael, Phys. Lett. B **322**, 239 (1994).
17. M. Davier, S. Eidelman, A. Höcker and Z. Zhang, Eur. Phys. J. C **31**, 503 (2003).
18. T. Blum, Phys. Rev. Lett. **91**, 052001 (2003).
19. T. Blum, Nucl. Phys. Proc. Suppl. **140**, 311 (2005).
20. L. Barkov et al., Nucl Phys. B **256**, 365 (1985).
21. T. Kinoshita, B. Nizic, Y. Okamoto, Phys. Rev. D **31**, 2108 (1985).
22. A. Casas, C. Lopez and F. J. Yndurain, Phys. Rev. D **32**, 736 (1985).
23. L. Martinovic and S. Dubnicka, Phys. Rev. D **42**, 884 (1990).
24. S. Eidelman and F. Jegerlehner, Zeit. f. Phys. C **67**, 585 (1995).
25. D. H. Brown and W. A. Worstell, Phys. Rev. D **54**, 3237 (1996).
26. R. Alemany, M. Davier and A. Höcker, Eur. Phys. J. C **2**, 123 (1998).
27. M. Davier and A. Höcker, Phys. Lett. B **419**, 419 (1998).
28. M. Davier and A. Höcker, Phys. Lett. B **435**, 427 (1998).
29. A. Höcker, talk at the 32nd International Conference on High Energy Physics, 2004, Beijing, China [hep-ph/0410081].
30. F. Jegerlehner, Nucl. Phys. Proc. Suppl. **131**, 213 (2004).
31. J. F. de Troconiz and F. J. Yndurain, Phys. Rev. D **71**, 073008 (2005).
32. R. R. Akhmetshin et al. [CMD-2 Collaboration], *Measurement of* $e^+e^- \to \pi^+\pi^-$ *cross-section with CMD-2 around ρ meson*, hep-ex/9904027.
33. S. I. Eidelman, Nucl. Phys. Proc. Suppl. **98**, 281 (2001).
34. R. R. Akhmetshin et al. [CMD-2 Collaboration], Phys. Lett. B **578**, 285 (2004); these results were recently updated to include data on $e^+e^- \to \pi^+\pi^-$ from the 1998 run, see the talk by I. Logoshenko at the Europhysics High Energy Conference, Lissbon, Portugal, July 2005.
35. A. Aloisio et al. [KLOE Collaboration], Eur. Phys. J. C **33**, S656 (2004); A. G. Denig [KLOE Collaboration], AIP Conf. Proc. **717**, 83 (2004) [hep-ex/0311012].
36. R. Barate et al. [ALEPH Collaboration], Eur. Phys. J. C **4**, 409 (1998).
37. S. Anderson et al. [CLEO Collaboration, Phys. Rev. D **61**, 112002 (2000).
38. K. Ackerstaff et al. [OPAL Collaboration], Eur. Phys. J. C **7**, 571 (1999).
39. M. N. Achasov et al. [SND Collaboration], *Study of the process* $e^+e^- \to \pi^+\pi^-$ *in the energy region* $400 < \sqrt{s} < 1000$ *MeV*, hep-ex/0506076.
40. S. Binner, K. Melnikov and J. H. Kühn, Phys. Lett. B **459**, 279 (1999).
41. M. Davier, S. Eidelman, A. Höcker and Z. Zhang, Phys. Lett. B **582**, 27 (2004).
42. S. G. Gorishny, A. L. Kataev and S. A. Larin, Phys. Lett. B **259**, 144 (1991); L.R. Surguladze and M. A. Samuel, Phys. Rev. Lett. **66**, 560 (1991).
43. K. G. Chetyrkin, A. H. Hoang, J. H. Kühn, M. Steinhauser and T. Teubner, Eur. Phys. J. C **2**, 137 (1998).
44. J. Z. Bai et al. [BES Collaboration], Phys. Rev. Lett. **84**, 594 (2000); Phys. Rev. Lett. **88**, 101802 (2002).
45. J. L. Siegrist et al. [Mark I Collaboration], Phys. Rev. D **26**, 969 (1982).
46. C. Bacci et al. [$\gamma\gamma$2 Collaboration], Phys. Lett. B **86**, 234 (1979).

47. Z. Jakubowsky et al. [Crystal Ball Collaboration], Zeit. f. Phys. C **40**, 49 (1988);
C. Edwards et al. [Crystal Ball Collaboration], SLAC-PUB-5160, 1990.

48. S. Dubinsky, A. Korchin, N. Merenkov, G. Pancheri and O. Shekhovtsova, Eur. Phys. J. C **40**, 41 (2005).

49. S. R. Amendolia ct al., Phys. Lett. B **138**, 454 (1984).

50. E.P. Solodov, in *Proc. of the e^+e^- Physics at Intermediate Energies Conference* ed. Diego Bettoni, eConf **C010430**, T03 (2001) [hep-ex/0107027].

51. B. Aubert et al. [BABAR Collaboration], Phys. Rev. D **71**, 052001 (2001).

52. V. Cirigliano, G. Ecker, H. Neufeld, JHEP **0208**, 002 (2002).

53. T. Kinoshita and A. Sirlin, Phys. Rev. **113**, 585 (1959).

54. H. Czyz and J. H. Kühn, Eur. Phys. J. C **18**, 497 (2001).

55. W. J. Marciano and A. Sirlin, Phys. Rev. Lett. **61**, 1815 (1988).

56. T. Kinoshita, J. Math. Phys. **3**, 650 (1962).

57. T. D. Lee and M. Nauenberg, Phys. Rev. **133**, 1549 (1964).

58. E. Braaten and C. S. Li, Phys. Rev. D **42**, 3888 (1990).

59. M. Davier, Nucl. Phys. Proc. Suppl. **131**, 123 (2004); for an update, see S. Schael et al. [ALEPH Collaboration], Phys. Rept. **421**, 191 (2005).

60. J. Schwinger, in *Particles, Sources and Fields*, vol. III, Perseus Books, Reading, Massachusetts, 1989. See also M. Drees and K. Hikasa, Phys. Lett. B **252**, 127 (1990), where the misprint in Schwinger's formula is corrected.

61. T. Krupovnickas, talk at PHENO 2005, Madison, Wisconsin, May 2005; http://pheno.physics.wisc.edu.

62. F. Guerrero and A. Pich, Phys. Lett. B **412**, 382 (1997).

63. J. Bijnens and P. Gosdzinsky, Phys. Lett. B **388**, 203 (1996).

64. S. Ghozzi and F. Jegerlehner, Phys. Lett. B **583**, 222 (2004).

65. M. Davier, S. Eidelman, A. Höcker and Z. Zhang, Eur. Phys. J. C **27**, 497 (2003).

66. B. Krause, Phys. Lett. B **390**, 392 (1997).

4 Electroweak Corrections to a_μ

4.1 Weak Corrections to a_μ

In this chapter we discuss electroweak corrections to the muon anomalous magnetic moment in the Standard Model. Their study started approximately thirty years ago, when the Standard Model was still raising to prominence.

At the one-loop level, there are three contributions to the electroweak correction to a_μ, Fig. 4.1; they appear due to exchanges of the W-boson, the Z-boson and the Higgs boson. Since the masses of all of these particles are large, compared to the muon mass, the structure of the one-loop electroweak correction is simple; apart from the coupling constants, it depends on the ratios $m^2/m^2_{Z,W,H}$. In addition, since the muon Yukawa coupling is small, the Higgs boson exchange is negligible. The one-loop result reads [1]

$$a_\mu^{\mathrm{ew},1l} = \frac{G_F m^2}{8\sqrt{2}\pi^2}$$

$$\times \left[\frac{10}{3} + \frac{1}{3}\left(-5 + (1 - 4\sin^2\theta_W)^2 \right) + \mathcal{O}\left(\frac{m^2}{m^2_{W,H}} \right) \right], \quad (4.1)$$

where the first term in square brackets in (4.1) refers to the W-boson and the second one to the Z-boson contribution to a_μ. Using $G_F = 1.16639(1) \times 10^{-5}$ GeV^2 and $\sin^2\theta_W = 0.224$, it evaluates to

$$a_\mu^{\mathrm{ew},1l} = 194.9 \times 10^{-11} . \quad (4.2)$$

The magnitude of the one-loop electroweak correction to a_μ, (4.2), suggests that at the current level of precision, the two-loop electroweak corrections to a_μ, suppressed by $\alpha/\pi \sim 2 \times 10^{-3}$, are not needed. However, as was

Fig. 4.1. Weak contributions to the muon anomalous magnetic moment

K. Melnikov and A. Vainshtein: *Theory of the Muon Anomalous Magnetic Moment*
STMP **216**, 89–102 (2006)

pointed out by Kukhto *et al.* [2], this estimate is too naive since the two-loop corrections are enhanced by large logarithms of the ratio of the Z boson mass to the muon mass, $\ln(m_Z/m) \sim 6.8$. Nevertheless, it is natural to expect that even such a logarithmic enhancement by itself would not lead to a significant two-loop effect since an estimate $a_\mu^{\mathrm{ew},2l} \sim a_\mu^{\mathrm{ew},1l}(\alpha/\pi)\ln(m_Z/m)$, leads to a correction to the muon anomalous magnetic moment of about 3×10^{-11}. Surprisingly, this estimate fails completely; explicit calculation of the two-loop electroweak corrections to a_μ [2, 3, 4, 5] shows that they change the one-loop result (4.2) by about -40×10^{-11}. Interestingly, those corrections are that large accidentally; the two-loop electroweak contribution to the muon magnetic anomaly can, approximately, be written as

$$a_\mu^{\mathrm{ew},2l} \sim -10 \ \left(\frac{\alpha}{\pi}\right) a_\mu^{\mathrm{ew},1l} \left(\ln \frac{m_Z}{m} + 1\right) , \qquad (4.3)$$

and the factor ten in the above equation appears since many "order one" contributions from individual diagrams add up coherently.

There is an additional peculiarity associated with the two-loop electroweak effects. Among various two-loop diagrams that contribute to the muon anomalous magnetic moment, there are diagrams that involve light quarks as intermediate particles. Some of those diagrams receive contributions from the region of small loop momenta where perturbative description of strong interactions breaks down and the use of quarks as degrees of freedom is unjustified; this situation is analogous to the hadronic vacuum polarization contribution to the muon anomalous magnetic moment discussed in Chap. 3.

There are two types of the light-quark diagrams where perturbative approach fails. The first one is the hadronic vacuum polarization contribution due to $\gamma - Z$ mixing. The contribution of the second type is due to graphs where a quark triangle subgraph formed by one axial and two vector currents is present. Such triangle diagrams are related to the axial anomaly of QCD and, for this reason, are sensitive to both the ultra-violet *and* the infra-red structure of the theory that is determined by hadronic interactions at low energies.

In the calculation of [3], the non-perturbative effects in diagrams with light quarks were modeled by introducing constituent quark masses. Being a reasonable and easy-to-implement way to account for non-perturbative effects, this method violates the chiral structure of hadronic interactions at low energies and, hence, introduces the theoretical uncertainty in the calculation. It is reasonable to expect that this theoretical uncertainty is comparable to the magnitude of the non-logarithmic term in (4.3). Note also, that even perturbative QCD effects in quark diagrams can be quite sizable; this happens because, numerically, $(\alpha_s/\pi)\ln(m_Z/m) \sim 1$, which implies that the three-loop mixed QCD-electroweak corrections to a_μ are *naturally* of the same magnitude as the two-loop electroweak correction *without* the logarithmic enhancement. Hence, we conclude that the full two-loop calculation of the electroweak effects can be justified and used if an only if, both perturbative and non-perturbative, QCD effects in two-loop electroweak corrections are better understood. Fortunately, a refined treatment of those effects recently became available [6, 7, 8]; we describe it in Chap. 5.

Our discussion of the two-loop electroweak corrections to the muon magnetic anomaly in the remainder of this chapter is restricted to the logarithmic approximation. We may do that because i) the computation of the logarithmically-enhanced two-loop electroweak corrections to the muon magnetic anomaly is relatively simple and, hence, can be discussed in a concise manner and ii) as follows from the complete calculation of the two-loop electroweak effects [3, 4, 5], the logarithmically-enhanced terms dominate the two-loop result. However, before discussing the two-loop electroweak corrections in the logarithmic approximation, we briefly explain how the complete calculation [3] was performed.

The calculation of the two-loop electroweak corrections to a_μ is plagued by technical difficulties; first, the number of diagrams is large ($\sim$1700 in the 't Hooft-Feynman gauge) and, second, these diagrams depend on a variety of particle masses. Those two facts make an analytic computation of a traditional pen-and-paper type impossible. Instead, one uses symbolic manipulation programs, such as FORM [9], to handle large algebraic expressions. In addition, the problem of computing two-loop diagrams that involve different masses is solved with the help of the asymptotic expansion of Feynman diagrams [10]. For further discussion of the calculation, the reader is referred to the original publication [3].

4.2 Two-loop Electroweak Effects: The Logarithmic Approximation

4.2.1 Effective Field Theory

In this section we describe the computation of the two-loop electroweak corrections to the muon anomalous magnetic moment in the logarithmic approximation, retaining terms $\mathcal{O}(G_F m^2 \alpha \ln(m_{W,Z}/m))$. Our discussion follows closely [7]. The calculation can be performed in an effective field theory, where all particles with masses comparable to or heavier than the weak scale $\sim m_Z \sim 100$ GeV, are integrated out. The resulting theory is the Fermi theory of weak interactions supplemented with the muon magnetic dipole operator. The degrees of freedom in the effective theory are leptons, photons and quarks with $m_q \ll m_Z$, i.e. all the quarks besides the top quark.

The effective theory at the scale μ_r such that $m \ll \mu_r \ll m_Z$ is described by the Lagrangian,

$$\mathcal{L}_{\text{eff}}(\mu_r) = -\frac{G_F}{2\sqrt{2}}\left(h(\mu_r)H(\mu_r) + \sum_i c^i(\mu_r)\mathcal{O}_i(\mu_r)\right), \qquad (4.4)$$

where the sum extends over dimension six four-fermion operators.

The operator $H(\mu_r)$ is the magnetic dipole operator

$$H(\mu_r) = \frac{m(\mu_r)}{16\pi^2}\, e(\mu_r)\left[F_{\mu\nu}\bar\mu\sigma^{\mu\nu}\mu\right]_{\mu_r}, \tag{4.5}$$

where $m(\mu_r)$ and $e(\mu_r) = -\sqrt{4\pi\alpha(\mu_r)}$ are the *running* (e.g. $\overline{\mathrm{MS}}$) muon mass and electric charge. These factors appear naturally in the Wilson coefficient of dimension 5 operator $F_{\mu\nu}\bar\mu\sigma^{\mu\nu}\mu$ because only Euclidean momenta larger than μ_r contribute to this coefficient; however, it is convenient to include those factors into the definition of the magnetic dipole operator H because, for example, the product $eF_{\alpha\beta}$ is μ_r-independent.

The electroweak contribution to the muon magnetic anomaly a_μ^{ew} is given by the Wilson coefficient $h(\mu_r)$ at the low normalization point $\mu_r = m$,

$$a_\mu^{\mathrm{ew}} = \frac{G_F m^2}{8\pi^2\sqrt{2}}\, h(m). \tag{4.6}$$

The one-loop contribution to a_μ^{ew}, (4.1), arises because of the Z and W exchanges; hence, it is determined by virtual momenta of order $m_{W,Z}$ and fixes the value of h at a high normalization point,

$$h(m_Z) = h^W + h^Z = \frac{10}{3} + \frac{1}{3}\left(-5\,(g_A^\mu)^2 + (g_V^\mu)^2\right). \tag{4.7}$$

Here $g_{A,V}^\mu$ are the axial and the vector couplings of the Z-boson to the muon

$$g_A^\mu = -1\,, \qquad g_V^\mu = -1 + 4\sin\theta_W^2\,. \tag{4.8}$$

Higher order corrections to the one-loop expressions (4.7) for $h^W(m_W)$ and $h^Z(m_Z)$ are of order α and *do not* contain large logarithms of the type $\log(m_Z/m)$ because at high normalization point only large virtual momenta $\sim m_Z$ contribute. The logarithmically enhanced electroweak corrections to the muon magnetic anomaly, $\sim \alpha\log(m_Z/m)$, appear because of the running of the Wilson coefficient $h(\mu_r)$ from $\mu_r = m_Z$ to $\mu_r = m$.

The running of $h(\mu_r)$ is governed by the renormalization group equation

$$\mu_r\frac{\mathrm{d}h(\mu_r)}{\mathrm{d}\mu_r} = -\frac{\alpha(\mu_r)}{2\pi}\left[\gamma_H h(\mu_r) + \sum_i \beta_i\, c^i(\mu_r)\right], \tag{4.9}$$

where γ_H and β_i are elements of the anomalous dimension matrix for operators H and $\mathcal{O}_i$. In particular, γ_H represents the anomalous dimension of the magnetic dipole operator H while β_i is determined by the mixing of the four-fermion operator $\mathcal{O}_i$ with H. In general, the anomalous dimensions are computed as series in the fine structure constant α. However, since we restrict ourselves to the two-loop corrections, γ_H and β_i are just numbers; we compute them below. To calculate $h(\mu_r)$ through $\mathcal{O}(\alpha^2)$, we integrate (4.9) over μ_r and approximate the Wilson coefficients and the coupling constant in the right-hand side of that equation by their values at $\mu_r = m_Z$, i.e. $h(\mu_r) = h(m_Z)$, $c^i(\mu_r) = c^i(m_Z)$, and $\alpha(\mu_r) = \alpha(m_Z)$. We obtain

$$h(\mu_r) = \left(1 + \gamma_H \frac{\alpha(m_Z)}{2\pi} \ln \frac{m_Z}{\mu_r}\right) h(m_Z) + \frac{\alpha(m_Z)}{2\pi} \ln \frac{m_Z}{\mu_r} \sum_i \beta_i\, c^i(m_Z) \,.$$

$$(4.10)$$

The dimension six four-fermion operators in the effective Lagrangian (4.4) originate from the Z and W exchanges between fermions. The exchange of the Z-boson between fermions f_1 and f_2 gives

$$\mathcal{L}_Z^{d=6}(m_Z) = -\frac{G_F}{4\sqrt{2}}\, j_\mu^Z\, j^{Z,\mu} \,, \qquad j_\mu^Z = \sum_f \left[g_V^f\, \bar{f}\gamma_\mu f + g_A^f\, \bar{f}\gamma_\mu\gamma_5 f\right], \quad (4.11)$$

where the vector and axial couplings are defined as

$$g_V^f = 2I_3^f - 4Q_f \sin^2\theta_W; \qquad g_A^f = 2I_3^f \,. \qquad (4.12)$$

Here I_3^f is the third component of the weak isospin of the fermion f and Q_f is its electric charge. For example, $I_3^{\mu,e,\tau} = -1/2$, $I_3^u = 1/2$, $I_3^d = -1/2$ and $Q_{e,\mu,\tau} = -1$, $Q_u = 2/3$, $Q_d = -1/3$.

Since the magnetic dipole operator H conserves parity, only vector-vector and axial-axial components of a generic four-fermion operator can mix with H. Therefore, the part of the $\mathcal{L}_Z^{d=6}$ Lagrangian, that is relevant for the computation of the muon anomalous magnetic moment, can be written as

$$\mathcal{L}_Z^{d=6,P=+1}(m_Z) = -\frac{G_F}{2\sqrt{2}} \sum_{f_1,f_2;\Gamma=V,A} c^{\Gamma;f}(m_Z)\mathcal{O}_{\Gamma;f_1,f_2} \,, \qquad (4.13)$$

where $c^{\Gamma;f_1,f_2}(m_Z) = g_\Gamma^{f_1} g_\Gamma^{f_2}$ and

$$\mathcal{O}_{V;f_1,f_2} = \frac{1}{2}\, \bar{f}_1\gamma^\mu f_1\, \bar{f}_2\gamma_\mu f_2 \,, \qquad \mathcal{O}_{A;f_1,f_2} = \frac{1}{2}\, \bar{f}_1\gamma^\mu\gamma_5 f_1\, \bar{f}_2\gamma_\mu\gamma_5 f_2 \,. \quad (4.14)$$

Finally, we point out that in order to mix with H, at least one of the fermions $f_{1,2}$ in a four-fermion operator $\mathcal{O}_{V(A);f_1,f_2}$ should be a muon. For this reason, the four-fermion operators produced by an exchange of the W-boson do not contribute. Indeed, the product of two charged currents containing muons involves a muon neutrino which does not interact with electromagnetic fields. Hence, it is not possible to construct a two-loop diagram in the effective field theory that contributes to the muon magnetic anomaly and involves any of the four-fermion operators that originate from the W exchanges. Note, however, that this becomes possible starting from the three-loop order. We conclude that (4.13, 4.14) contain all operators and Wilson coefficients, required to compute the two-loop electroweak corrections to the muon magnetic anomaly in the logarithmic approximation; what is missing are the anomalous dimensions γ_H and β_i. We discuss their computation in what follows.

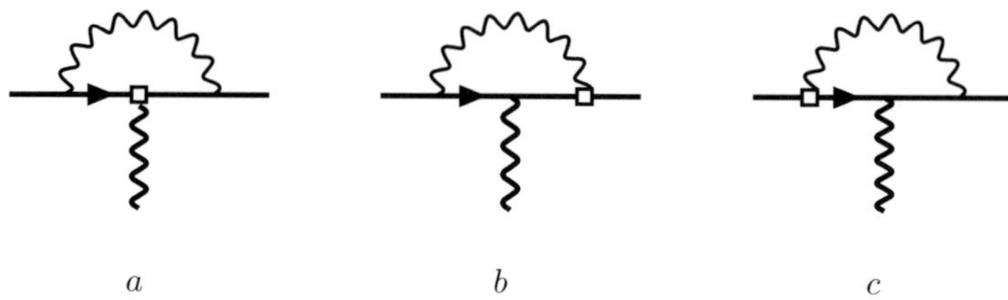

Fig. 4.2. Renormalization of the dipole magnetic operator

4.2.2 Anomalous Dimension and Mixing of Effective Operators

Calculation of the anomalous dimension of the magnetic dipole operator H is straightforward. Consider the matrix element $\langle\mu|H(m_Z)|\mu\gamma\rangle$ in the one-loop approximation. The soft photon in the final state imitates the external magnetic field. Three contributing diagrams are shown in Fig. 4.2. It is convenient to use the Landau gauge (3.95) for the computation, since the muon wave function renormalization in this gauge is finite. The matrix element reads

$$\langle\mu|H(m_Z)|\mu\gamma\rangle = \frac{em(m_Z)}{8\pi^2}\,\epsilon^\mu\bar{u}_{p'}\left[i\sigma_{\mu\nu}q^\nu + e^2\left(\Gamma_\mu^{(a)} + \Gamma_\mu^{(b)} + \Gamma_\mu^{(c)}\right)\right]u_p\,,$$

$$(4.15)$$

where ϵ and q are the polarization vector and the momentum of the external photon and

$$\Gamma_\mu^{(a)} = \int \frac{\mathrm{d}^4k}{(2\pi)^4}\,\frac{\gamma^\alpha(\hat{p}'+\hat{k})\sigma_{\mu\nu}q^\nu(\hat{p}+\hat{k})\gamma^\beta}{k^2(k^2+2pk)(k^2+2p'k)}\,\mathcal{P}_{\alpha\beta}(k)\,,$$

$$\Gamma_\mu^{(b)} = \int \frac{\mathrm{d}^4k}{(2\pi)^4}\,\frac{\sigma_{\alpha\beta}k^\beta(\hat{p}'+\hat{k})\gamma_\mu(\hat{p}+\hat{k})\gamma^\alpha}{k^2(k^2+2pk)(k^2+2p'k)}\,, \qquad (4.16)$$

$$\Gamma_\mu^{(c)} = -\int \frac{\mathrm{d}^4k}{(2\pi)^4}\,\frac{\gamma^\alpha(\hat{p}'+\hat{k})\gamma_\mu(\hat{p}+\hat{k})\sigma_{\alpha\beta}k^\beta}{k^2(k^2+2pk)(k^2+2p'k)}\,.$$

The integration over the loop momentum k in (4.16) is restricted to the Euclidean range $\mu_r < k < m_Z$. Since we work in the logarithmic approximation, we may choose $\mu_r \gg m$ and neglect the muon mass m in $\Gamma_\mu^{(a,b,c)}$.

We begin with the computation of $\Gamma_\mu^{(a)}$ that corresponds to the diagram Fig. 4.2a. This diagram is easy to compute, since it is explicitly proportional to first power of q, the momentum of the external photon. As a consequence, we neglect p and p' in $\Gamma_\mu^{(a)}$ and obtain

$$\Gamma_\mu^{(a)} = \int \frac{\mathrm{d}^4k}{(2\pi)^4}\,\frac{\gamma^\alpha\hat{k}\sigma_{\mu\nu}q^\nu\hat{k}\gamma^\beta}{k^6}\,\mathcal{P}_{\alpha\beta}(k)\,. \qquad (4.17)$$

Averaging over directions of the vector k and using $\gamma^\rho\sigma_{\mu\nu}\gamma_\rho = 0$, we arrive at

$$\Gamma_\mu^{(a)} = -\int \frac{\mathrm{d}^4 k}{(2\pi)^4} \frac{\sigma_{\mu\nu} q^\nu}{k^4} = -\frac{i}{8\pi^2} \sigma_{\mu\nu} q^\nu \int_{\mu_r}^{m_Z} \frac{dk}{k} = -\frac{i}{8\pi^2} \sigma_{\mu\nu} q^\nu \ln \frac{m_Z}{\mu_r} \ .$$

$$(4.18)$$

A somewhat more involved algebra is required to compute $\Gamma_\mu^{(b,c)}$, that correspond to diagrams b and c in Fig. 4.2. To extract the logarithmic contribution, we have to expand $\Gamma_\mu^{(b,c)}$ in (4.16) to first order in p and p'. Upon doing so and averaging over directions of the momentum k, we obtain

$$\Gamma_\mu^{(b)} = \Gamma_\mu^{(c)} = -\frac{i}{4\pi^2} \sigma_{\mu\nu} q^\nu \ln \frac{m_Z}{\mu_r} \ . \tag{4.19}$$

The final step in deriving a relation between the magnetic dipole operators normalized at different points, $H(m_Z)$ and $H(\mu_r)$, involves expressing $m(m_Z)$, that appears in the definition of $H(m_Z)$, through $m(\mu_r)$. With the logarithmic accuracy, the relation between the two masses reads

$$m(m_Z) = m(\mu_r) \left(1 - 3 \frac{\alpha}{2\pi} \ln \frac{m_Z}{\mu_r}\right) . \tag{4.20}$$

Using (4.15, 4.18, 4.19, 4.20) to collect all the ingredients, we arrive at

$$\langle \mu | H(m_Z) | \mu\gamma \rangle = \left[1 - 8 \frac{\alpha}{2\pi} \ln \frac{m_Z}{\mu_r}\right] \langle \mu | H(\mu_r) | \mu\gamma \rangle \ . \tag{4.21}$$

Comparing this result to (4.10), we find the anomalous dimension of the magnetic dipole operator,

$$\gamma_H = -8 \ . \tag{4.22}$$

4.2.3 Mixing of Four-fermion Operators; Triangle Diagrams

Having discussed the radiative corrections to the matrix element of the magnetic dipole operator, we turn our attention to four-fermion operators. There are different cases that have to be considered. The simplest one involves the matrix element of the operator $\mathcal{O}_{V;\mu f}$, where f is different from μ. The corresponding two-loop diagram is shown in Fig. 4.3. It is convenient to write the matrix element $\langle \mu | \mathcal{O}_{V;\mu f} | \mu\gamma \rangle$ in the following way [1]

$$\langle \mu | \mathcal{O}_{V;\mu f}(m_Z) | \mu\gamma \rangle = i\, e^3 Q_f N_f\, \epsilon^\mu\, \bar{u}_{p'} \Gamma_\mu^V u_p \ , \tag{4.23}$$

$$\Gamma_\mu^V = \int \frac{\mathrm{d}^4 k}{(2\pi)^4} \frac{\gamma^\alpha (\hat{p}' + \hat{k} + m)\gamma_\mu \left(\hat{p} + \hat{k} + m\right)\gamma^\beta}{(k^2 + 2p'k)(k^2 + 2pk)} \mathcal{P}_{\alpha\beta}(k)\Pi(k^2) \ ,$$

---

[1] Note that when discussing the mixing of four-fermion operators with the magnetic dipole operator, we compute the contribution of $\mathcal{O}_{\Gamma;\mu f}$, rather than the contribution of $\mathcal{O}_{\Gamma;\mu f} + \mathcal{O}_{\Gamma;f\mu}$. The missing symmetry factor is restored in the final result for the two-loop electroweak correction to the muon anomalous magnetic moment, (4.47).

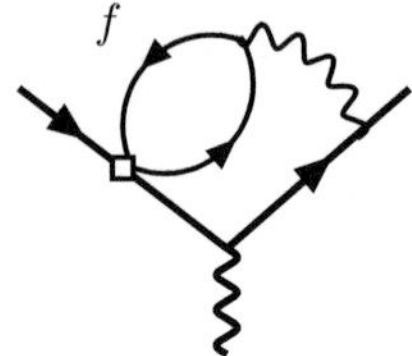

Fig. 4.3. Mixing of the operator $\mathcal{O}_{V;\mu f}$ with the magnetic dipole operator. The symmetric diagram is not shown

where N_f accounts for color degrees of freedom ($N_f = 3(1)$ for quarks (leptons)) and $\Pi(k^2)$ is the polarization operator

$$i\Pi(k^2)k^2\mathcal{P}_{\alpha\beta} = \int \mathrm{d}^4x\, \mathrm{e}^{ikx}\langle 0|T\{j_\alpha(x)j_\beta(0)\}|0\rangle \;, \tag{4.24}$$

for the current $j_\alpha = \bar{f}\gamma_\alpha f$. We have multiplied (4.23) by a factor 2, to account for the symmetric diagram where the operator $\mathcal{O}_{V;\mu f}$ is inserted into the outgoing muon line.

The matrix element in (4.23) can be easily computed because the function $\Pi(k^2)$ is the correlator of two vector currents for which the unsubtracted dispersion representation reads

$$\Pi(k^2) = \frac{1}{12\pi^2} \int\limits_{4m_f^2}^{\infty} \frac{\mathrm{d}s\,\rho(s)}{s - k^2} \;, \quad \rho(s) = \sqrt{1 - \frac{4m_f^2}{s}}\left(1 + \frac{2m_f^2}{s}\right). \tag{4.25}$$

Substituting (4.25) into (4.23) and changing the order of integration, we obtain

$$\Gamma_\mu^V = \frac{1}{12\pi^2} \int\limits_{4m_f^2}^{\infty} \mathrm{d}s\,\rho(s)\,\mathcal{T}_\mu(s), \tag{4.26}$$

$$\mathcal{T}_\mu(s) = -\int \frac{\mathrm{d}^4k}{(2\pi)^4}\, \frac{\gamma^\alpha\left(\hat{p}' + \hat{k} + m\right)\gamma_\mu\left(\hat{p} + \hat{k} + m\right)\gamma^\beta}{(k^2 + 2p'k)(k^2 + 2pk)(k^2 - s)}\,\mathcal{P}_{\alpha\beta}(k).$$

It is easy to see that $\mathcal{T}_\mu(s)$ is proportional to the one-loop correction to the matrix element of the vector current due to an exchange of a "photon" with the mass s. The exchange of such "photon" generates a correction to the anomalous magnetic moment of the muon that scales as $1/s$ for $s \gg m^2$. Using (2.67), we obtain

$$\mathcal{T}_\mu(s) = \frac{m}{24\pi^2 s}\,\sigma_{\mu\nu}q^\nu \;. \tag{4.27}$$

A subsequent integration over s leads to a logarithmically enhanced correction to the muon anomalous magnetic moment. Using (4.26, 4.23), we arrive at

$$\langle\mu|\mathcal{O}_{V;\mu f}(m_Z)|\mu\gamma\rangle = \frac{4}{9}\,N_f Q_f\,\frac{\alpha}{2\pi}\,\ln\frac{m_Z}{\mu_r}\,\langle\mu|H(\mu_r)|\mu\gamma\rangle\,, \tag{4.28}$$

which implies that the mixing coefficient $\beta_{V;\mu f}$ for $f\neq\mu$ is

$$\beta_{V;\mu f} = \frac{4}{9}\,N_f Q_f\,. \tag{4.29}$$

Next, we consider the mixing of the operator $\mathcal{O}_{A,\mu f}$, for $f\neq\mu$ with the magnetic dipole operator. The corresponding diagram is shown in Fig. 4.4.

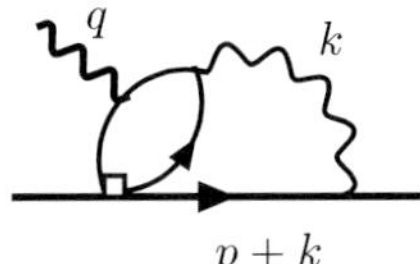

Fig. 4.4. Mixing of the operator $\mathcal{O}_{A;\mu f}$ with the magnetic dipole operator

The fermion-loop sub-graph of that diagram involves the correlator of two electromagnetic vector currents and an axial current; it is the famous fermion triangle associated with the anomaly in the divergence of the axial current. For this reason it was thoroughly studied in the literature. We require this sub-graph in a particular kinematics, where the momentum q of one of the vector currents is small and we only retain terms linear in q. In that case, the triangle amplitude can be represented as the matrix element of the correlator of the electromagnetic current J_μ and the axial current $A_\nu = \bar{f}\gamma_\nu\gamma_5 f$, between the soft photon with momentum q and the vacuum,

$$T_{\mu\nu} = \langle 0|\hat{T}_{\mu\nu}|\gamma(q)\rangle\,, \quad \hat{T}_{\mu\nu} = i\int \mathrm{d}^4x\,\mathrm{e}^{ikx}T\{J_\mu(x)A_\nu(0)\}\,. \tag{4.30}$$

Physically, the limit of small photon momentum q means that we probe the muon magnetic moment with the homogeneous magnetic field.

The particular kinematic limit described above allows us to use Schwinger operator methods to compute the correlator in (4.30). The leading contribution to that correlator comes from the fermion loop, with fermion propagators computed in the external constant electromagnetic field. Since we are interested in the matrix element of $\hat{T}_{\mu\nu}$ between a single soft photon and the vacuum, we only require the correlator through first order in the external field. In the fixed-point gauge $x^\mu A_\mu = 0$, the vector potential A_μ can be expressed through the electromagnetic field-strength tensor $F_{\mu\nu} = \partial_\mu A_\nu - \partial_\nu A_\mu$; the relation reads $A_\mu(x) = -1/2 F_{\mu\nu}x^\nu + \mathcal{O}(x^2)$. We use $1/(i\hat{\partial}-m-eQ_f A_\mu(x)\gamma^\mu)$ for the fermion propagator, expand in the external electromagnetic field through first order and arrive at the following expression for the fermion propagator [12, 13]

$$S(p) = \frac{1}{\hat{p} - m_f} - \frac{1}{(p^2 - m_f^2)^2}\, eQ_f\, \tilde{F}_{\rho\delta}\left(p^\rho\gamma^\delta - \frac{i}{2}\,m_f\,\sigma^{\rho\delta}\right)\gamma_5 + \mathcal{O}(F^2)\,,$$

$$(4.31)$$

where $\tilde{F}_{\rho\delta} = (1/2)\epsilon_{\rho\delta\alpha\beta}F^{\alpha\beta}$ is the dual electromagnetic field-strength tensor. We are now in position to compute the contribution of a fermion f to $\hat{T}_{\mu\nu}$,

$$\hat{T}_{\mu\nu} = i\int\frac{\mathrm{d}^4p}{(2\pi)^4}\,\mathrm{Tr}\left[\gamma_\mu S(p+k)\gamma_\nu\gamma_5 S(p)\right].$$

$$(4.32)$$

At zeroth order in $F_{\alpha\beta}$, the integral over p vanishes while at first order in $F_{\alpha\beta}$ we obtain

$$\hat{T}_{\mu\nu} = -2ieQ_f^2 N_f\int\frac{\mathrm{d}^4p}{(2\pi)^4}\,\frac{\mathrm{Tr}\left[\gamma_\mu(\hat{p}+\hat{k}+m_f)\gamma_\nu\gamma_5\tilde{F}_{\rho\delta}\left(p^\rho\gamma^\delta - \frac{i}{2}\,m_f\,\sigma^{\rho\delta}\right)\gamma_5\right]}{[(p+k)^2 - m_f^2](p^2 - m_f^2)^2}\,.$$

$$(4.33)$$

The operator nature of $\hat{T}_{\mu\nu}$ is reflected in the presence of the electromagnetic field-strength tensor $F_{\alpha\beta}$ in (4.33). Introducing Feynman parameters, computing the trace and integrating over p, we arrive at

$$\hat{T}_{\mu\nu} = -\frac{eQ_f^2 N_f}{4\pi^2}\int_0^1\mathrm{d}\xi\,\frac{\xi(1-\xi)\left(k_\mu k^\rho\tilde{F}_{\rho\nu} + k_\nu k^\rho\tilde{F}_{\rho\mu}\right) - m_f^2\tilde{F}_{\mu\nu}}{m_f^2 - \xi(1-\xi)k^2}\,.$$

$$(4.34)$$

While (4.34) looks legitimate, in reality it is not because it does not satisfy the vector current conservation condition $k^\mu\hat{T}_{\mu\nu} = 0$. We restore the conservation of the vector current by adding a polynomial counterterm to (4.34); this counterterm can be constructed from the contribution to $\hat{T}_{\mu\nu}$ of an infinitely heavy Pauli-Villars regulator field

$$\hat{T}_{\mu\nu}^{\mathrm{phys}} = \hat{T}_{\mu\nu}(m_f) - \lim_{m_f\to\infty}\hat{T}_{\mu\nu}(m_f) = \hat{T}_{\mu\nu}(m_f) - \frac{eQ_f^2}{4\pi^2}\tilde{F}_{\mu\nu}\,,$$

$$(4.35)$$

where we used (4.34) to find $T_{\mu\nu}(m_f)$ in the limit $m_f\to\infty$. In this way we obtain

$$\hat{T}_{\mu\nu}^{\mathrm{phys}} = -\frac{eQ_f^2 N_f}{4\pi^2}\left(-k^2\tilde{F}_{\mu\nu} + k_\mu k^\rho\tilde{F}_{\rho\nu} + k_\nu k^\rho\tilde{F}_{\rho\mu}\right)\int_0^1\mathrm{d}\xi\,\frac{\xi(1-\xi)}{m_f^2 - \xi(1-\xi)k^2}$$

$$= \frac{eQ_f^2 N_f}{4\pi^2\, k^2}\left(-k^2\tilde{F}_{\mu\nu} + k_\mu k^\rho\tilde{F}_{\rho\nu} + k_\nu k^\rho\tilde{F}_{\rho\mu}\right)\left(1 + \frac{2m_f^2}{\beta k^2}\ln\frac{\beta+1}{\beta-1}\right).$$

$$(4.36)$$

In (4.36), we use $\beta = \sqrt{1 - 4m_f^2/k^2}$. In what follows, we omit the superscript in the tensor $T_{\mu\nu}$ that we used in (4.36) to denote properly regulated physical expression.

Contracting $\hat{T}_{\mu\nu}$ in (4.36) with k_μ, we observe that $k^\mu \hat{T}_{\mu\nu} = 0$ which implies that the conservation of the vector current is maintained. On the other hand, the divergence of the axial current does not vanish even in the chiral $m_f = 0$ limit, as can be seen from the following expression

$$k^\nu \hat{T}_{\mu\nu} = \frac{eQ_f^2 N_f}{2\pi^2}\, k^\rho \tilde{F}_{\rho\mu}\left(1 + \frac{2m_f^2}{\beta k^2}\ln\frac{\beta+1}{\beta-1}\right). \tag{4.37}$$

This is in accord with the operator equation for the divergence of the axial current

$$\partial^\mu A_\mu = -\frac{\alpha Q_f^2 N_f}{2\pi}\, F^{\alpha\beta}\tilde{F}_{\alpha\beta} + 2im_f\,\bar{f}\gamma_5 f\,, \tag{4.38}$$

where the first term represents the axial anomaly.

We can use (4.38, 4.37) to compute the fermion triangle diagram where the axial current is substituted by the pseudoscalar current $P = \bar{f}\gamma_5 f$. The first term at the right-hand side of (4.37) matches the first term at the right-hand side of (4.38); matching the remaining terms in these equations leads to

$$\hat{T}_\mu^P = i\int d^4x\, e^{ikx} T\{J_\mu(x)P(0)\} = m_f\,\frac{eQ_f^2 N_f}{2\pi^2}\,\frac{k^\rho \tilde{F}_{\rho\mu}}{\beta k^2}\ln\frac{\beta+1}{\beta-1}\,. \tag{4.39}$$

We will need this result in what follows. In addition, we will require the result for the fermion triangle where the axial vector current is substituted by the scalar current $S = \bar{f}f$. We compute it along the lines described above and obtain

$$\hat{T}_\mu^S = i\int d^4x\, e^{ikx} T\{J_\mu(x)S(0)\} = -im_f\,\frac{eQ_f^2 N_f}{2\pi^2}\,\frac{k^\rho F_{\rho\mu}}{k^2}\left[\frac{1+\beta^2}{2\beta}\ln\frac{\beta+1}{\beta-1} - 1\right].$$
$$\tag{4.40}$$

We now have all the ingredients that are needed to discuss the mixing of the remaining four-fermion operators with the magnetic dipole operator.

First, we finalize the computation of the mixing of the operator $\mathcal{O}_{A;\mu f}$ for $f \neq \mu$, with the magnetic dipole operator, Fig. 4.4. The fermion triangle subgraph is described by the amplitude $T_{\mu\nu}$, computed above. We use this amplitude to write the full matrix element

$$\langle\mu|\mathcal{O}_{A;\mu f}(m_Z)|\mu\gamma\rangle = ie^2\int\frac{d^4k}{(2\pi)^4}\,\frac{\bar{u}_{p'}\gamma^\mu\left(\hat{p}+\hat{k}+m\right)\gamma^\nu\gamma_5 u_p}{k^2(k^2+2pk)}\,T_{\mu\nu}(k)\,, \tag{4.41}$$

where we introduced a factor of two to account for a similar diagram where the virtual photon couples to the incoming muon. Since we are only interested in the logarithmically enhanced corrections, we require the amplitude $T_{\mu\nu}$, (4.36), in the limit $-k^2 \gg m^2$. Expanding the integrand in (4.41) through first order in p and m and integrating over k we derive

$$\langle \mu | \mathcal{O}_{A;\mu f}(m_Z) | \mu \gamma \rangle = -6\, N_f Q_f^2 \frac{\alpha}{2\pi} \ln \frac{m_Z}{\mu_r}\, \langle \mu | H(\mu_r) | \mu \gamma \rangle \,, \qquad (4.42)$$

which translates into the mixing coefficient

$$\beta_{A;\mu f} = -6\, N_f Q_f^2 \,, \qquad f \neq \mu \,. \qquad (4.43)$$

Finally, we have to compute the mixing of the operators $\mathcal{O}_{V,A;\mu\mu}$ with the magnetic dipole operator. In variance with the operators considered previously, there are two possibilities for the fermion pairing; μ and $\bar{\mu}$ from either the same or different currents can be Wick-contracted.

When μ and $\bar{\mu}$ from the same current are contracted, the results can be read off from (4.29, 4.43). To this end, we use $Q_f \to Q_\mu = -1$ and $N_f \to N_\mu = 1$; we also multiply the results by a combinatorial factor two due to two ways of pairing.

When muons from different currents are contracted, the calculation can be simplified using the Fiertz transformation; it reads

$$
\begin{aligned}
\bar{\psi}_1 \gamma_\mu \psi_2 \bar{\psi}_3 \gamma_\mu \psi_4 =& \frac{1}{2}\, \bar{\psi}_1 \gamma_\mu \psi_4 \bar{\psi}_3 \gamma_\mu \psi_2 + \frac{1}{2}\, \bar{\psi}_1 \gamma_\mu \gamma_5 \psi_4 \bar{\psi}_3 \gamma_\mu \gamma_5 \psi_2 \\
& - \bar{\psi}_1 \psi_4 \bar{\psi}_3 \psi_2 + \bar{\psi}_1 \gamma_5 \psi_4 \bar{\psi}_3 \gamma_5 \psi_2, \\
\bar{\psi}_1 \gamma_\mu \gamma_5 \psi_2 \bar{\psi}_3 \gamma_\mu \gamma_5 \psi_4 =& \frac{1}{2}\, \bar{\psi}_1 \gamma_\mu \gamma_5 \psi_4 \bar{\psi}_3 \gamma_\mu \gamma_5 \psi_2 + \frac{1}{2}\, \bar{\psi}_1 \gamma_\mu \psi_4 \bar{\psi}_3 \gamma_\mu \psi_2 \\
& + \bar{\psi}_1 \psi_4 \bar{\psi}_3 \psi_2 - \bar{\psi}_1 \gamma_5 \psi_4 \bar{\psi}_3 \gamma_5 \psi_2.
\end{aligned}
\qquad (4.44)
$$

Operators in (4.44) that involve products of vector and axial currents can be treated in the same way as what has been discussed above. Terms with scalar and pseudoscalar currents in (4.44) lead to additional contributions to the matrix elements of the operators $\mathcal{O}_{V(A);\mu\mu}$. Since insertions of scalar and pseudoscalar operators in the diagram Fig. 4.3 vanish in the Landau gauge, we only have to consider diagrams where scalar and pseudoscalar operators are inserted into fermion triangles. Those contributions read

$$
\begin{aligned}
\Delta^{S+P} \langle \mu | \mathcal{O}_{V;\mu\mu} | \mu \gamma \rangle &= -\Delta^{S+P} \langle \mu | \mathcal{O}_{A;\mu\mu} | \mu \gamma \rangle \\
&= 2ie^2 \int \frac{\mathrm{d}^4 k}{(2\pi)^4} \frac{1}{(k^2)^2} \, \bar{u}_{p'} \gamma^\mu \hat{k} (T_\mu^S - \gamma_5 T_\mu^P) u_p \\
&= 4 \cdot \frac{\alpha}{2\pi} \ln \frac{m_Z}{\mu_r} \, \langle \mu | H | \mu \gamma \rangle.
\end{aligned}
\qquad (4.45)
$$

where we used (4.40, 4.39) for T_μ^S and T_μ^P at $-k^2 \gg m^2$. We combine all the ingredients and derive the mixing coefficients for the operators $\mathcal{O}_{V,A;\mu\mu}$ with the magnetic dipole operator

$$\beta_{V;\mu\mu} = -\frac{4}{9} \cdot 3 - 6 + 4 = -\frac{10}{3} \,, \qquad \beta_{A;\mu\mu} = -\frac{4}{9} - 6 \cdot 3 - 4 = -\frac{202}{9} \,. \qquad (4.46)$$

4.2.4 The Two-loop Electroweak Correction

The full two-loop electroweak correction to the muon anomalous magnetic moment in the logarithmic approximation is given by the value of the Wilson coefficient h at the normalization point $\mu_r = m$, (4.6). Using the results for entries of the anomalous dimension matrix, it is straightforward to obtain $h(m)$. The only subtlety is that the running of $h(\mu_r)$, described by (4.10), changes once a particle threshold is passed; in other words, once the renormalization scale μ_r becomes smaller than the mass of the fermion f, this fermion decouples and stops to participate in the running. As the consequence, for muon and electron components, the running continues all the way down to $\mu_r = m$ but, for heavier fermions, it terminates at $\mu_r = m_f$. The entries of the anomalous dimension matrix can be found in (4.22, 4.29, 4.43, 4.46). We obtain

$$
a_\mu^{\mathrm{ew},2l} = \frac{G_F m^2}{8\pi^2 \sqrt{2}} \left(\frac{\alpha}{\pi}\right) \left\{ -\left(\frac{215}{9} + \frac{31}{9}(g_V^\mu)^2\right) \ln \frac{m_Z}{m} \right.
$$
$$
\left. + \sum_{f \neq e,\mu,t} \left(6 g_A^f N_f Q_f^2 + \frac{4}{9} g_V^\mu g_V^f N_f Q_f\right) \ln \frac{m_Z}{m_f} \right\}.
$$

(4.47)

In (4.47) the first term accounts for the muon and electron contributions. Note that in deriving (4.47) we have accounted for the symmetry factor related to the appearance of the operators $\mathcal{O}_{\Gamma;\mu f}$ and $\mathcal{O}_{\Gamma;f\mu}$ in the effective Lagrangian (4.4), where appropriate.

We can now check that the logarithmically enhanced terms indeed strongly dominate the two-loop electroweak correction to the muon magnetic anomaly. Numerically, the two-loop electroweak correction evaluates to $\sim -40 \times 10^{-11}$ [3]. Evaluating (4.47) with $m_u = m_d = m_s = 300$ MeV, we find that the two-loop electroweak corrections to a_μ in the logarithmic approximation are $\sim -33 \times 10^{-11}$.

Given the fact that the two-loop electroweak corrections are significant, one may wonder about the possible impact of three-loop logarithmically enhanced corrections to the muon magnetic anomaly. Such corrections are discussed in [7, 14]. The results of these analyses show that the three-loop electroweak corrections are extremely small. For example, if the two-loop result, (4.47) is written through the fine structure constant evaluated at the scale $\mu = m$, the three-loop leading logarithmic correction to a_μ is 0.4×10^{-11} and, for this reason, can be safely neglected at the current level of precision.

Finally, we note that the two-loop electroweak correction, (4.47), depends upon the light quark masses. Such a dependence is unphysical and may signal that some calculations discussed in this chapter fail to correctly account for the effects of strong interactions. The next chapter is dedicated to the analysis of hadronic effects in electroweak corrections. After that, we give the full result for the electroweak contribution to the muon anomalous magnetic moment that we use to derive the Standard Model value of a_μ, in Chap. 7.

References

1. W. A. Bardeen, R. Gastmans and B. Lautrup, Nucl. Phys. B **46**, 315 (1972);
   G. Altarelli, N. Cabibbo and L. Maiani, Phys. Lett. B **40**, 415 (1972);
   R. Jackiw and S. Weinberg, Phys. Rev. D **5**, 2473 (1972);
   I. Bars and M. Yoshimura, Phys. Rev. D **6**, 374 (1972);
   M. Fujikawa, B.W. Lee and A.I. Sanda, Phys. Rev. D **6**, 2923 (1972).
2. T. V. Kukhto, E. A. Kuraev, A. Schiller and Z. K. Silagadze, Nucl. Phys. B **371**, 567 (1992).
3. A. Czarnecki, B. Krause and W. J. Marciano, Phys. Rev. Lett. **76**, 3267 (1996).
4. S. Heinemeyer, D. Stockinger and G. Weiglein, Nucl. Phys. B **699**, 103 (2004).
5. T. Gribouk and A. Czarnecki, Phys. Rev. D **72**, 053016 (2005).
6. S. Peris, M. Perrottet and E. de Rafael, Phys. Lett. B **355**, 523 (1995).
7. A. Czarnecki, W. J. Marciano and A. Vainshtein, Phys. Rev. D **67**, 073006 (2003).
8. M. Knecht, S. Peris, M. Perrottet and E. de Rafael, JHEP **11**, 003 (2002).
9. J. A. Vermaseren, *New features of FORM*, math-ph/0010025.
10. V.A. Smirnov, *Applied Asymptotic Expansions In Momenta And Masses*, Springer, Berlin (2002).
11. S. L. Adler, Phys. Rev. **177**, 2426 (1969);
    L. Rosenberg, Phys. Rev. **129**, 2786 (1963).
12. V. A. Novikov, M. A. Shifman, A. I. Vainshtein and V. I. Zakharov, Fortsch. Phys. **32**, 585 (1984).
13. E.V. Shuryak and A.I. Vainshtein, Nucl. Phys. B **201**, 141 (1982);
14. G. Degrassi and G. F. Giudice, Phys. Rev. D **58**, 053007 (1998).

5 Strong Interaction Effects in Electroweak Corrections

5.1 Anomalous Quark Triangles: Effects of Strong Interactions

We have seen in the previous chapter that the calculation of the two-loop electroweak corrections to the muon anomalous magnetic moment involves correlators of an axial current with two vector currents. Studies of such correlators in late sixties lead to the discovery of the anomaly of the axial current [1, 2]; implications of this phenomenon include computation of the decay amplitude for $\pi_0 \to \gamma\gamma$, 't Hooft consistency condition [3], solution of the U(1) problem etc. In case of the muon magnetic moment, such "anomalous" current correlators are important as well.

To illustrate that, we return to the calculation of the mixing of the four-quark operator $\mathcal{O}_{A;\mu f}$, (4.14), with the magnetic dipole operator $H(\mu_r)$, (4.5), described by the graph in Fig. 4.4. Such graphs contain a correlator of an axial current with two vector currents. We have seen that integration over momentum of the virtual photon in diagram Fig. 4.4 produces a logarithm $\ln(m_Z/m_{\max})$, where $m_{\max} = \max\{m, m_q\}$ with m_q being the mass of the quark in the triangle subgraph. When the quark is heavy, $m_q \gg \Lambda_{\mathrm{QCD}} \gtrsim m$, its treatment in perturbation theory is justified since loop integrals are cut off in the infra-red at the mass of the heavy quark. However, for light, e.g. u and d, quarks the logarithmic integration extends all the way down to very small virtualities where effects of strong interactions invalidate perturbative calculations.

The simplest modification of the perturbative computation of the diagrams Fig. 4.4, that accounts for some effects of strong interactions, is the use of constituent masses for light quarks, whereby $m_u \sim m_d \sim m_s \sim 300$ MeV. In this case, even for light quarks, the integration over loop momentum is cut off at the scale $\sim \Lambda_{\mathrm{QCD}}$ which is of the order of the pion mass. Such an approach has been taken in the original calculation of the two-loop electroweak corrections to the muon anomalous magnetic moment [4].

While approximating quark masses by their constituent values is likely to be reasonable in as much as numerical results are concerned, it is not well-founded theoretically since it violates chiral structure of QCD. Therefore, it is interesting to ask if a consistent account of strong interaction effects in

K. Melnikov and A. Vainshtein: *Theory of the Muon Anomalous Magnetic Moment*
STMP **216**, 103–120 (2006)
© Springer-Verlag Berlin Heidelberg 2006

electroweak corrections is possible. Such a possibility was first pointed out in [5, 6]. The forthcoming discussion follows [7], where some flaws in the analyses of [5, 6] are identified and corrected.

The centerpiece of our discussion is the correlator of the axial and two vector currents. For the computation of the muon magnetic anomaly, we require this correlator in a particular kinematic limit where one of the vector currents, that describes the external magnetic field, carries vanishing momentum q. We need to study such correlator through first order in q; all higher order terms can be neglected for our purposes.

Because we are interested in the correlator of three currents in a particular kinematic limit, it is convenient to work with the operator $\hat{T}_{\mu\nu}$ representing the T-product of two "hard" currents,

$$\hat{T}_{\mu\nu} = \sum_f 2I_3^f \, \hat{T}_{\mu\nu}^f = i \int \mathrm{d}^4 x\, e^{ikx} T\left\{ J_\mu(x) J_\nu^5(0) \right\},\qquad (5.1)$$

where

$$J_\mu = \sum_f Q_f \bar{f}\gamma_\mu f\,, \qquad J_\nu^5 = \sum_f 2I_3^f \, \bar{f}\gamma_\nu\gamma_5 f\,,\qquad (5.2)$$

and $\hat{T}_{\mu\nu}^f$ is defined in (4.30) of the previous section. In (5.2), J_μ is the usual electromagnetic current and J_ν^5 describes the axial part of the $Z\bar{f}f$ coupling.

The anomalous fermion triangle Fig. 5.1 corresponds to the matrix element of $\hat{T}_{\mu\nu}$ between the vacuum and the on-shell photon with momentum q and polarization ϵ,

$$T_{\mu\nu} = \sum_f 2I_3^f T_{\mu\nu}^f = \langle 0|\hat{T}_{\mu\nu}|\gamma(q)\rangle\,.\qquad (5.3)$$

We are interested in $T_{\mu\nu}$ in the limit of small photon momentum $q \to 0$. Thanks to gauge invariance, the expansion of $T_{\mu\nu}$ in powers of q starts with a linear term that appears in the form of the field-strength tensor of the electromagnetic field $f_{\alpha\beta} = q_\alpha\epsilon_\beta - q_\beta\epsilon_\alpha$. Since under parity transformation $T_{\mu\nu}$ transforms as a pseudo-tensor, we may write it as a linear combination of three Lorentz structures $\tilde{f}_{\mu\nu}$, $k_\mu k^\alpha \tilde{f}_{\alpha\nu}$ and $k_\nu k^\alpha \tilde{f}_{\alpha\mu}$, where $\tilde{f}_{\alpha\beta} = (1/2)\epsilon_{\alpha\beta\rho\sigma} f^{\rho\sigma}$ is the dual of the electromagnetic field-strength tensor. It is possible to reduce the number of independent Lorentz structures using the transversality of $T_{\mu\nu}$ with respect to the Lorentz index of the vector current, $k^\mu T_{\mu\nu} = 0$. We may choose two independent Lorentz structures in such a way that one of them is transversal and the other is longitudinal with respect to the Lorentz index of the axial current ν. We therefore write

$$T_{\mu\nu} = \frac{-i|e|}{4\pi^2}\left[w_{\mathrm{T}}(k^2)\{ -k^2\tilde{f}_{\mu\nu} + k_\mu k^\alpha \tilde{f}_{\alpha\nu} - k_\nu k^\alpha \tilde{f}_{\alpha\mu}\} + w_{\mathrm{L}}(k^2) k_\nu k^\alpha \tilde{f}_{\alpha\mu}\right].$$
$$(5.4)$$

The parametrization in terms of the functions $w_{\mathrm{L,T}}$ is convenient since, once those functions are computed, the correction to the anomalous magnetic moment is obtained from the integral

$$
a_\mu^{\mathrm{ew},\Delta} = 2\sqrt{2}G_F m^2 \left(\frac{\alpha}{\pi}\right) \int \frac{\mathrm{d}^4 k}{(2\pi)^4} \frac{i}{k^2 + 2kp}
$$
$$
\times \left[\frac{1}{3}\left(1 + \frac{2(kp)^2}{k^2 m^2}\right)\left(w_{\mathrm{L}} - \frac{m_Z^2}{m_Z^2 - k^2} w_{\mathrm{T}}\right) + \frac{m_Z^2}{m_Z^2 - k^2} w_{\mathrm{T}}\right].
$$

$$(5.5)$$

Here p denotes the four-momentum of the on-shell muon, $p^2 = m^2$. To estimate $a_\mu^{\mathrm{ew},\Delta}$ with the logarithmic accuracy, a much simpler expression is sufficient

$$
a_\mu^{\mathrm{ew},\Delta} \approx \frac{G_F m^2}{8\pi^2 \sqrt{2}} \left(\frac{\alpha}{\pi}\right) \int_0^\infty \mathrm{d}K^2 \left(w_{\mathrm{L}} + \frac{m_Z^2}{m_Z^2 + K^2} w_{\mathrm{T}}\right).
$$

$$(5.6)$$

Here we use Euclidean momentum K, i.e. $K^2 = -k^2$. Note that in case of a heavy, $m_f \gg m$, fermion contribution to $w_{\mathrm{L,T}}$, (5.6) can be used to derive $a_\mu^{\mathrm{ew},\Delta}$ with a power accuracy, $\mathcal{O}(m^2/m_f^2)$.

As we see, the structure functions $w_{\mathrm{L,T}}$ are important for computing electroweak corrections to the muon anomalous magnetic moment. As it turns out, computation of those functions requires careful consideration of both perturbative and non-perturbative effects of strong interactions. The following discussion is dedicated to such an analysis.

5.1.1 Perturbative Calculations

At lowest order in the strong coupling constant, the functions $w_{\mathrm{L,T}}$ are obtained by calculating the one-loop triangle diagram, Fig. 5.1. For a general kinematics, this diagram was first computed in [8]; the result was later simplified [9] in the limit of the vanishing photon momentum q. In Chap. 4 we derived the expression for $\hat{T}_{\mu\nu}^f$, (4.36); comparing it to (5.4), we find

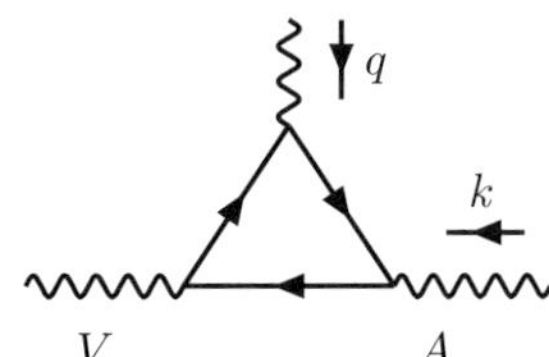

Fig. 5.1. Quark triangle diagram that determines the functions $w_{\mathrm{L,T}}$ in the one-loop approximation. Diagram with opposite direction of the fermion lines, is not shown

$$w_{\mathrm{L}}^f = 2w_{\mathrm{T}}^f = 2N_f Q_f^2 \int_0^1 \mathrm{d}\xi \, \frac{\xi(1-\xi)}{\xi(1-\xi)K^2 + m_f^2}$$

$$= \frac{2N_f Q_f^2}{K^2} \left(1 - \frac{2m_f^2}{\beta K^2} \ln \frac{\beta+1}{\beta-1} \right) , \tag{5.7}$$

where $N_f = 3(1)$ for quarks (leptons) and $\beta = \sqrt{1 + 4m^2/K^2}$. From (5.7), it is easy to derive asymptotic expressions for $w_{\mathrm{L,T}}$ in the limit of small and large K^2,

$$w_{\mathrm{L}}^f = 2w_{\mathrm{T}}^f = 2N_f Q_f^2 \times \begin{cases} \dfrac{1}{K^2} \left(1 - \dfrac{2m_f^2}{K^2} \ln \dfrac{K^2}{m_f^2} \right) , & K^2 \gg m_f^2 ; \\[4mm] \dfrac{1}{6m_f^2} \left(1 - \dfrac{K^2}{5m_f^2} \right) , & K^2 \ll m_f^2 . \end{cases} \tag{5.8}$$

Recall, that the anomaly in the divergence of the axial current is seen from the fact that $w_L^f \neq 0$ for $m_f = 0$.

At large K^2, the structure functions decrease as $1/K^2$; then, for the transversal structure function w_{T}^f the integral in (5.6) converges thanks to the presence of $m_Z^2/(K^2 + m_Z^2)$ factor, while for the longitudinal structure function w_{L}^f the result appears to be divergent. This divergence, however, disappears, once the sum over *quarks and leptons in a given generation* is performed. Indeed, retaining $\mathcal{O}(K^{-2})$ terms, we obtain the large-K^2 asymptotics of the longitudinal structure function

$$w_{\mathrm{L}} = \sum_{\mathrm{generation}} 2I_3^f w_{\mathrm{L}}^f \quad \longrightarrow \quad \frac{4}{K^2} \sum_{\mathrm{generation}} I_3^f N_f Q_f^2 = 0 . \tag{5.9}$$

The last equality in (5.9) follows from the relation

$$\sum_{\mathrm{generation}} I_3^f N_f Q_f^2 = 0 , \tag{5.10}$$

which is nothing but the anomaly cancelation condition in the Standard Model, required for the internal consistency of the theory.

If masses of all fermions in a generation are the same, the contribution of such generation to w_{L} vanishes. When fermions in a given generation have substantially different masses, the corrections to the anomalous magnetic moment contain logarithms of the mass ratios, e.g. $\ln(m_t/m_\tau)$ for the third and $\ln(m_c/m_\mu)$ for the second generation, etc.

For the contribution of the transversal function w_{T} to the muon magnetic anomaly, the situation is similar up to a clarification. Thanks to the factor $m_Z^2/(K^2 + m_Z^2)$, that multiplies w_{T} in (5.6), the contribution of the transversal structure function to a_μ depends on the hierarchy between fermion masses in a given generation and the mass of the Z-boson. Clearly, this issue only concerns the third generation; for the first two generations, the factor $m_Z^2/(K^2 + m_Z^2)$ in (5.6) can be set to one, with the accuracy m_f^2/m_Z^2. Then, similar to the longitudinal structure function, the dependence on $\ln(m_Z)$ cancels out due to (5.10) and logarithms of the fermion mass ratios appear in the final result.

As we see, the disappearance of $\ln m_Z$ terms in $a_\mu^{\mathrm{ew},\Delta}$ is intimately related to the cancelation between lepton and quark contributions at large values of the loop momentum $K^2 \gg m_f^2$. Since quark contributions may be modified by gluon exchanges, it is appropriate to ask if this cancelation between lepton and quark loops holds beyond the one-loop approximation to $w_{\mathrm{L,T}}$?

In general, gluon exchanges may produce $\mathcal{O}(\alpha_s)$ corrections to quark contribution to the structure functions. Even if such corrections are present in w_{L}, the integral in (5.6) still diverges as $\int \mathrm{d}K^2 \alpha_s(K^2)/K^2$, in spite of the logarithmic decrease of the strong coupling constant $\alpha_s(K^2) \propto 1/\log(K^2/\Lambda_{\mathrm{QCD}}^2)$. The Adler-Bardeen theorem [2] protects w_{L} from such corrections; the theorem states that, in the chiral limit $m_q = 0$, there are no QCD corrections to w_{L}. Hence, for large values of the loop momentum, $K^2 \gg m_q^2$, the QCD corrections to w_{L} have to be mass-suppressed $\delta w_{\mathrm{L}}/w_{\mathrm{L}} \sim \alpha_s(K^2) m_q^2/K^2$, which is sufficient for the ultra-violet convergence. On the other hand, the heavy quark loop integration converges in the infra-red, so that the major contribution to w_{L} comes from the virtualities $K^2 \sim m_q^2 \gg \Lambda_{\mathrm{QCD}}^2$.

Thus, for light quarks, the Adler–Bardeen theorem implies that QCD corrections to w_{L} vanish in the chiral limit. For heavy quarks, QCD corrections to w_{L} do not vanish; they are proportional to $\alpha_s(m_q)/\pi$. However, relative corrections are even smaller since leading terms are enhanced by logarithms of fermion mass ratios but those logarithms do no appear in $\mathcal{O}(\alpha_s)$ corrections.

In case of the transversal structure function w_{T}, the problem of the ultra-violet divergence of the integral in (5.6) is absent, thanks to the factor $m_Z^2/(K^2+m_Z^2)$. Nevertheless, if $\mathcal{O}(\alpha_s)$ corrections to w_{T} are present at large $K^2 \gg m_q^2$, they lead to a logarithmic sensitivity to the Z-boson mass, $\sim \log\log m_Z$; fortunately, as was recently shown by Vainshtein [10], this does not happen since, in the chiral limit, the transversal structure function w_{T} does not receive perturbative QCD corrections.

To prove this statement, we note that the amplitude $T_{\mu\nu}$ possesses certain symmetry properties under the permutation of indices μ and ν that refer to vector and axial currents, respectively. This symmetry follows from the fact that, in the chiral limit $m_q = 0$, the Dirac matrix γ_5 from the axial current can be anti-commuted through even number of γ-matrices to the vector current. Changing also the direction of the momentum, $k \to -k$, which does not affect $T_{\mu\nu}$, we observe that the amplitude $T_{\mu\nu}$ is symmetric under the permutation of μ and ν.

This symmetry seems to be in contradiction with the perturbative result, (5.4, 5.7), evaluated at $m_f = 0$. The calculation of $\hat{T}_{\mu\nu}$, described in the previous chapter, demonstrates, however, that this symmetry is broken by the regularization, cf. (4.34, 4.36) for $m_f = 0$. To avoid the necessity to deal with the regularization, we apply the symmetry argument to the imaginary part of $T_{\mu\nu}$. Then, the following equality holds

$$\mathrm{Im}\, T_{\mu\nu} = \mathrm{Im}\, T_{\nu\mu} \,. \tag{5.11}$$

When (5.11) is combined with the parametrization for $T_{\mu\nu}$ (5.4), we obtain the system of equations for the functions $w_{\mathrm{L,T}}$:

$$k^2 \mathrm{Im}[w_{\mathrm{T}}(k^2)] = 0, \qquad 2\,\mathrm{Im}[w_{\mathrm{T}}(k^2)] = \mathrm{Im}[w_{\mathrm{L}}(k^2)] \,. \tag{5.12}$$

Once the imaginary parts of the functions $w_{\mathrm{L,T}}$ are calculated, we restore the real parts of these functions using dispersion representation

$$w_{\mathrm{L,T}}(k^2) = \frac{1}{\pi} \int\limits_0^\infty \mathrm{d}s\, \frac{\mathrm{Im}[w_{\mathrm{L,T}}(s)]}{s - k^2 - i0} \,. \tag{5.13}$$

Note, that no subtraction term in (5.13) is allowed on dimensional grounds.

A possible solution to (5.12) is $\mathrm{Im}\,[w_{\mathrm{L,T}}] = 0$, but this implies that $w_{\mathrm{L,T}} = 0$, due to (5.13). The only nontrivial solution is then

$$2\,\mathrm{Im}[w_{\mathrm{T}}(k^2)] = \mathrm{Im}[w_{\mathrm{L}}(k^2)] = c\delta(k^2) \,, \tag{5.14}$$

and the constant c can be fixed from the one-loop computation since, as follows from the Adler-Bardeen theorem, the longitudinal structure function is not renormalized by higher order perturbative QCD effects. We derive

$$c = 2\pi N_f Q_f^2 \,. \tag{5.15}$$

Hence, in the chiral $m_f = 0$ limit,

$$w_{\mathrm{L}}^f = 2w_{\mathrm{T}}^f = \frac{2N_f Q_f^2}{K^2} \,, \tag{5.16}$$

to all orders in the strong coupling constant [2, 10].

5.1.2 Non-perturbative Effects and the OPE

While perturbatively, to all orders in the strong coupling constant, the longitudinal and transversal functions $w_{\mathrm{L,T}}$ for massless quarks differ only by an overall factor, this can not hold at the non-perturbative level. Indeed, in perturbation theory, these functions have poles at $k^2 = 0$. Provided that these poles survive the transition from perturbation theory to the real world with hadrons as degrees of freedom, they should describe propagation of massless hadrons. Pions and η mesons are the only non-strange massless particles in the chiral limit $m_u = m_d = m_s = 0$. The isospin numbers of the combination of the longitudinal structure functions

$$w_{\mathrm{L}}^{I=1} = w_{\mathrm{L}}^u - w_{\mathrm{L}}^d = \frac{2}{K^2} \ , \tag{5.17}$$

coincide with the isospin quantum numbers of a neutral pion. Therefore, we may interpret the pole at $K^2 = 0$ in (5.17) as corresponding to a massless pion that is emitted from the axial current and then decays into two photons. This duality between perturbative result, (5.17), and the possibility to interpret it using hadronic degrees of freedom, is the basis of the 't Hooft consistency condition [3] which states that *non-perturbative* corrections to $w_{\mathrm{L}}^{I=1}$, (5.17), are absent. There is yet another combination of the longitudinal structure functions, with the SU(3) quantum numbers of the η meson, $w_{\mathrm{L}}^u + w_{\mathrm{L}}^d - 2w_{\mathrm{L}}^s$, that does not receive non-perturbative corrections in the chiral limit. Because, in the chiral limit, there are no other neutral strangeless massless mesons besides π_0 and η, all other combinations of w_{L}^q as well as all transversal structure functions w_{T}^q should not have singularities at $K^2 = 0$; hence, they *must* receive non-perturbative corrections. Below we discuss these non-perturbative effects using the method of operator product expansion (OPE), described in Sect. 3.2.2.

Consider the correlator $T_{\mu\nu}$, (5.1), in the limit of large Euclidean momentum $K^2 \to \infty$. In that limit, the T-product of the vector and axial currents obeys an expansion in powers of $1/K^2$

$$\hat{T}_{\mu\nu} = \sum_i c_{\mu\nu}^{(i)\alpha_1\ldots\alpha_i} \mathcal{O}_{\alpha_1\ldots\alpha_i}^{(i)} \ . \tag{5.18}$$

Operators $\mathcal{O}_{\alpha_1\ldots\alpha_i}^{(i)}$ are ordered by their mass dimension. If the mass dimension of an operator is d_i, its Wilson coefficient scales like

$$c_{\mu\nu}^{(i)\alpha_1\ldots\alpha_i} \propto \frac{1}{K^{d_i-2}} \ . \tag{5.19}$$

The amplitude $T_{\mu\nu}$ is then derived from the matrix elements of the operators $\mathcal{O}_{\alpha_1\ldots\alpha_i}$ between the vacuum and the soft photon with the momentum q

$$T_{\mu\nu} = \sum c_{\mu\nu}^{(i)\alpha_1\ldots\alpha_i} \langle 0|\mathcal{O}_{\alpha_1\ldots\alpha_i}^{(i)}|\gamma(q)\rangle \ . \tag{5.20}$$

As we mentioned earlier, the T-product of the axial and vector currents $\hat{T}_{\mu\nu}$ is a pseudotensor. If we require the coefficients $c_{\mu\nu}^{(i)\alpha_1\ldots\alpha_i}$ to be tensors under Lorentz transformations, the operators $\mathcal{O}_{\alpha_1\ldots\alpha_i}$ have to be pseudotensors. Their matrix elements $\langle 0|\mathcal{O}_{\alpha_1\ldots\alpha_i}|\gamma(q)\rangle$ are linear in $\tilde{f}_{\alpha\beta}$; as a consequence, the operators $\mathcal{O}_{\alpha_1\ldots\alpha_i}$ should transform as $(1,0) - (0,1)$ under Lorentz transformations. Since we work to first order in the momentum of the soft photon, only operators $\mathcal{O}_{\alpha\beta}^{(i)}$ with two antisymmetric indices contribute. It is convenient to parametrize their matrix elements as follows

$$\langle 0|\mathcal{O}^{(i)}_{\alpha\beta}|\gamma(q)\rangle = -\frac{i|e|}{4\pi^2}\,\kappa_i \tilde{f}_{\alpha\beta}\;. \tag{5.21}$$

The coefficients κ_i are determined by non-perturbative dynamics; in general, they are not computable. Parametrically, these constants scale as $\kappa_i \sim \Lambda_{\rm QCD}^{d_i-2}$, where d_i is the mass dimension of the operator $\mathcal{O}_i$.

We can simplify the notation for the Wilson coefficients using the transversality of $T_{\mu\nu}$ with respect to the Lorentz index μ of the vector current. We write

$$\hat{T}_{\mu\nu} = \sum_i \left\{ c_{\rm T}^{(i)}(k^2)\left(-k^2 g_\mu^\alpha g_\nu^\beta + k_\mu k^\alpha g_\nu^\beta - k_\nu k^\alpha g_\mu^\beta\right) + c_{\rm L}^{(i)}(k^2)k_\nu k^\alpha g_\mu^\beta \right\} \mathcal{O}^{(i)}_{\alpha\beta}\;. \tag{5.22}$$

Finally, in terms of the Wilson coefficients $c_{\rm L,T}$, the functions $w_{\rm L,T}$ are written as

$$w_{\rm L,T}(k^2) = \sum_i c_{\rm L,T}^{(i)}(k^2)\kappa_i\;, \tag{5.23}$$

where the sum is over all operators that contribute to the OPE of the amplitude $T_{\mu\nu}$.

To compute the operator product expansion for $\hat{T}_{\mu\nu}$, we follow the strategy described in Sect. 3.2.2. To this end, we consider Feynman diagrams that describe a correlator of the vector and axial currents with a possible emission of a photon and study different routes for the large momentum k to travel from vector to axial current. Integrating over the loop momenta associated with these "hard" lines, we obtain the Wilson coefficients; the operators $\mathcal{O}^{(i)}_{\alpha\beta}$ are then read off from the soft lines in a diagram.

The leading contribution to the functions $w_{\rm L,T}$ is obtained from the triangle diagram Fig. 5.1 where *all* lines are considered to be hard. The operator that is produced by this momentum configuration is the dual of the field-strength tensor of the electromagnetic field $\mathcal{O}_F = |e|/(4\pi^2)\tilde{F}_{\alpha\beta}$ of the mass dimension two; its matrix element between the vacuum and the single photon state is

$$\langle 0|\frac{|e|}{4\pi^2}\,\tilde{F}_{\alpha\beta}|\gamma(q)\rangle = -\frac{i|e|}{4\pi^2}\,\tilde{f}_{\alpha\beta}\;, \tag{5.24}$$

so that $\kappa_F = 1$ (cf. (5.21)).

To study how operators of higher dimension contribute to the matrix element $T_{\mu\nu}$, we have to distinguish between hard and soft gluon exchanges. First, consider the case when all gluon exchanges in the diagrams that describe the correlator of vector and axial currents are soft and the large momentum k flows through the quark line that connects the axial and vector vertices. Such diagrams generate bilinear fermion operators. The Wilson coefficients of these operators are obtained by considering diagrams of the Compton scattering type in a soft external field, see Fig. 5.2.

It is easy to see that Wilson coefficients $c_{\rm L,T}$ for such diagrams vanish in the chiral limit. This follows from the symmetry of the matrix element $T_{\mu\nu}$

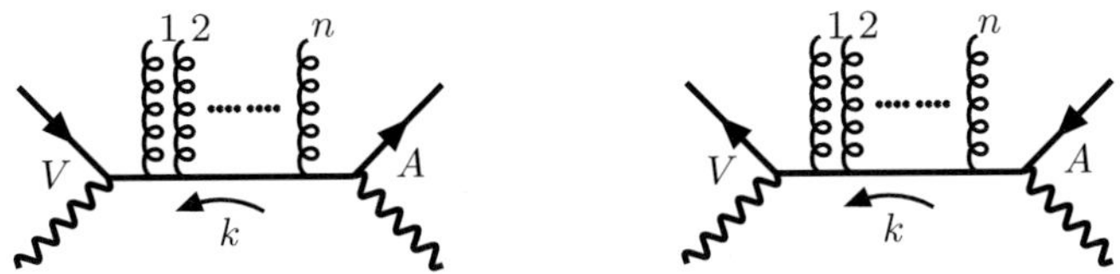

Fig. 5.2. Compton scattering amplitude in the external color field

with respect to μ and ν permutations; the argument is identical to the one, used earlier to prove that perturbative corrections to the structure functions $w_{\mathrm{L,T}}$ are absent in the chiral limit.

Since there is a continuous fermion line in the diagram Fig. 5.2, we can commute γ_5 from the axial current to the vector current; changing the direction of the momentum k, we derive

$$T_{\mu\nu}(k) = T_{\nu\mu}(-k) \,. \tag{5.25}$$

Using (5.22) and the antisymmetry of $\tilde{f}_{\alpha\beta}$, we see that the Wilson coefficients of the operators produced by diagrams Figs.5.2 satisfy the following equations

$$c_{\mathrm{T}}^i(k^2) = -c_{\mathrm{T}}^i(k^2) \,, \qquad 2c_{\mathrm{T}}^i(k^2) = c_{\mathrm{L}}^i(k^2) \,. \tag{5.26}$$

It follows that $c_{\mathrm{T}}^i(k^2) = c_{\mathrm{L}}^i(k^2) = 0$, for all fermion bilinears, independent of the number of soft gluons.

The same argument applies to diagrams shown in Fig. 5.3 that produce operators of the form $\bar{f}\Gamma f\tilde{F}$ and $\tilde{F}G^a G^a$ with $G_{\mu\nu}^a$ being the field-strength tensor of the gluon field. Since those diagrams do not require regularization, our argument applies and their Wilson coefficients vanish in the chiral limit.

Another way to arrive at the same conclusion is to use the flavor $\mathrm{U}(3)_L \times \mathrm{U}(3)_R$ symmetry of the massless theory for non-anomalous Wilson coefficients. To contribute to the operator product expansion of $T_{\mu\nu}$, the operator should be non-singlet with respect to both $\mathrm{U}(3)$ groups; hence, it must contain both left- and right-handed quarks. But, for the operators $\bar{f}\Gamma f\tilde{F}$ this is not possible since along the fermion line the chirality is conserved and the operators $\tilde{F}G^a G^a$ contain no fermions.

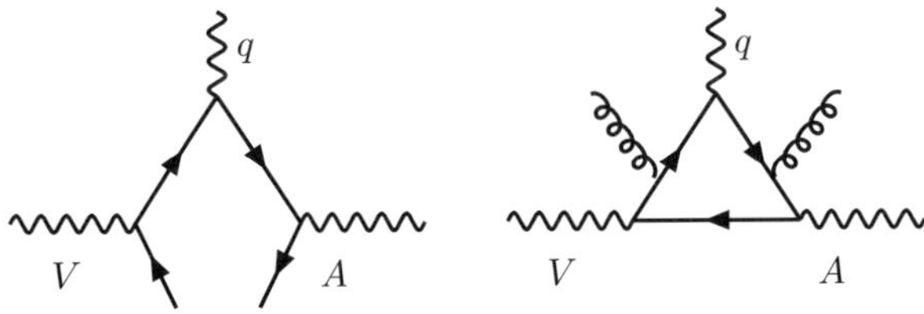

Fig. 5.3. Other non-perturbative contributions to $T_{\mu\nu}$ that vanish in the chiral limit

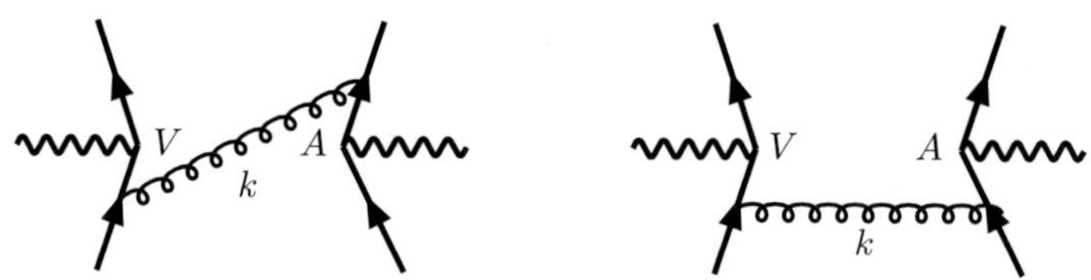

Fig. 5.4. Sample diagrams for the leading non-perturbative contribution to $T_{\mu\nu}$

Therefore, to get a non-vanishing contribution to $T_{\mu\nu}$ in the chiral limit, we need to consider diagrams with no fermion line connecting the vector current and the axial current; since the large momentum still has to flow from one vertex to the other, this requires an exchange of at least one *hard gluon* [6]. The leading contribution comes from diagrams with a single hard gluon exchange; they are shown in Fig. 5.4. These diagrams lead to a non-perturbative contribution given by dimension six four-fermion operators.

Consider first the non-perturbative contribution generated by a quark q. The part of $\hat{T}_{\mu\nu}$ due to diagrams shown in Fig. 5.4 reads

$$\Delta T^q_{\mu\nu} = -\frac{8\pi\alpha_s Q_q}{k^6}\,\bar{q}\,t^a\big(\gamma_\alpha\hat{k}\gamma_\mu - \gamma_\mu\hat{k}\gamma_\alpha\big)q \otimes \bar{q}\,t^a\big(\gamma_\nu\hat{k}\gamma^\alpha - \gamma^\alpha\hat{k}\gamma_\nu\big)\gamma_5 q\ , \quad (5.27)$$

where t^a are the generators of the SU(3) color group. Note that $\Delta T^q_{\mu\nu}$ is transverse with respect to Lorentz indices of *both* vector and axial currents; this immediately implies that *only* the transversal function $w_{\rm T}$ receives the corresponding non-perturbative contribution. This is an illustration of a general situation with *non-perturbative corrections* to the longitudinal structure function $w_{\rm L}$ mentioned earlier – similar to perturbative corrections, the non-perturbative corrections to $w_{\rm L}$ are absent in the chiral limit,[1] in accord with the 't Hooft consistency condition for the axial anomaly [3].

To simplify (5.27), we use

$$\gamma_\alpha\hat{k}\gamma_\mu - \gamma_\mu\hat{k}\gamma_\alpha = -2i\epsilon_{\alpha\mu\rho\beta}k^\rho\gamma^\beta\gamma_5\ , \quad (5.28)$$

and rewrite $\Delta T^q_{\mu\nu}$ as

$$\Delta T^q_{\mu\nu} = -\frac{16\pi\alpha_s Q_q}{k^6}\big(-k^2 g_{\mu\alpha}g_{\nu\beta} + k_\mu k_\alpha g_{\nu\beta} - k_\nu k_\alpha g_{\mu\beta}\big)$$
$$\times \big(\bar{q}\,t^a\gamma^\alpha\gamma_5 f \otimes \bar{q}\,t^a\gamma^\beta q - \alpha \leftrightarrow \beta\big)\ , \quad (5.29)$$

where only antisymmetric part of the matrix element with respect to indices α,β is shown. We introduce the dimension six operator

$$\mathcal{O}^6_{\alpha\beta} = \bar{q}\,t^a\gamma_\alpha\gamma_5 q \otimes \bar{q}\,t^a\gamma_\beta q - \alpha \leftrightarrow \beta\ , \quad (5.30)$$

---

[1] Note, that the flavor singlet longitudinal function $w_{\rm L}^u + w_{\rm L}^d + w_{\rm L}^s$ *does* receive the non-perturbative contribution. It appears at the level where the anomalous triangle with axial current and two gluons enters the OPE coefficient.

and, using (5.29), find its coefficient function

$$c_{\mathrm{T}}^{6,q}(k^2) = -\frac{8Q_q\pi\alpha_s(K^2)}{K^6} \ . \tag{5.31}$$

To determine the non-perturbative contribution to the transversal function $w_{\mathrm{T}}(K^2)$, we have to estimate the matrix element $\langle 0|\mathcal{O}_{\alpha\beta}^6|\gamma(q)\rangle$, which is easy to do in the vacuum saturation approximation. Using

$$\langle 0|\bar{q}_\sigma^i q_\rho^j|0\rangle = \frac{\delta^{ij}\delta_{\sigma\rho}}{4N_c}\langle 0|\bar{q}q|0\rangle \ , \tag{5.32}$$

with $i(j)$ and $\alpha(\beta)$ being SU(3) color and Lorentz indices respectively, we derive

$$\langle 0|\mathcal{O}_{\alpha\beta}^6|\gamma\rangle = -\frac{N_c^2-1}{2N_c^2}\langle 0|\bar{q}q|0\rangle \ \langle 0|i\bar{q}\sigma_{\alpha\beta}\gamma_5 q|\gamma\rangle \ . \tag{5.33}$$

The matrix element $\langle 0|i\bar{q}\sigma_{\alpha\beta}\gamma_5 q|\gamma\rangle$ was discussed in [11] where nucleon magnetic moments were studied using QCD sum rules. The numerical results are presented in terms of the quantity called *magnetic susceptibility* of the quark condensate $\chi = -(350 \pm 50 \text{ MeV})^{-2}$ defined as

$$\langle 0|i\bar{q}\sigma_{\alpha\beta}\gamma_5 q|\gamma\rangle = -i|e|\,\chi\,Q_q\langle 0|\bar{q}q|0\rangle\,\tilde{f}_{\alpha\beta} \ . \tag{5.34}$$

With the help of (5.21, 5.33, 5.34), we obtain

$$\kappa_6^q = -2\pi^2\chi\,Q_q\,\frac{N_c^2-1}{N_c^2}\,\langle 0|\bar{q}q|0\rangle^2 \ . \tag{5.35}$$

Combining (5.23, 5.31) and (5.35), we derive the leading non-perturbative contribution to the function $w_{\mathrm{T}}(K^2)$ for the first generation in the chiral limit

$$\Delta^{(6)}w_{\mathrm{T}}[u,d] = \sum_{q=u,d} 2I_3^q\Delta^{(6)}w_{\mathrm{T}}^q = \frac{256\pi^3\alpha_s(K^2)}{9K^6}\sum_{q=u,d}I_3^q Q_q^2\chi\,\langle 0|q\bar{q}|0\rangle^2 \ . \tag{5.36}$$

For numerical estimates, we use $\langle 0|q\bar{q}|0\rangle = -(235 \text{ MeV})^3$. We then find

$$\Delta^{(6)}w_{\mathrm{T}}[u,d] = -\alpha_s(K^2)\frac{(0.7 \text{ GeV})^4}{K^6} \tag{5.37}$$

Equation (5.37) illustrates a striking difference between the longitudinal and the transversal structure functions in the chiral limit. The longitudinal structure function is known *exactly* in the chiral limit,

$$w_{\mathrm{L}}[u,d] = \frac{2}{K^2} \ ; \tag{5.38}$$

there are *no perturbative or non-perturbative corrections to this result.* On the other hand, the transversal function receives non-perturbative corrections; its large-K^2 asymptotics is given by

$$w_{\mathrm{T}}[u,d] = \frac{1}{K^2}\left[1 - \alpha_s(K^2)\frac{(0.7\ \mathrm{GeV})^4}{K^4}\right].\tag{5.39}$$

The results discussed so far are derived in the chiral limit, which is not realized in Nature. It is therefore important to study the consequences of the fact that the chiral symmetry is explicitly violated by non-zero quark masses. To this end, we repeat the OPE analysis of the product of the axial and vector currents $\hat{T}_{\mu\nu}$, allowing for non-zero quark masses. It is then easy to see that diagrams with soft gluon exchanges shown in Fig. 5.2 give non-vanishing contribution to the OPE once non-zero quark masses are allowed. Hence, the leading non-perturbative contribution to $T_{\mu\nu}^q$ for a massive quark is obtained from the diagram Fig. 5.2 in the absence of soft gluon exchanges. The non-perturbative operator, generated in this way is $-i\bar{q}\sigma_{\alpha\beta}\gamma_5 q$; its Wilson coefficients are

$$c_{\mathrm{L}}^q = 2c_{\mathrm{T}}^q = \frac{4Q_q m_q}{K^4}.\tag{5.40}$$

The corresponding contribution to the longitudinal and transversal functions $w_{\mathrm{L,T}}$ is

$$\Delta^{(3)}w_{\mathrm{L}} = 2\Delta^{(3)}w_{\mathrm{T}} = \frac{8}{K^4}\sum_q I_f^3 Q_q m_q \kappa_q,\tag{5.41}$$

where κ_q is related to the magnetic susceptibility χ

$$\kappa_q = -4\pi^2 Q_q\,\chi\,\langle 0|\bar{q}q|0\rangle.\tag{5.42}$$

Using the Gell-Mann–Oakes–Renner relation [12] $(m_u+m_d)\langle 0|\bar{q}q|0\rangle = F_\pi^2 m_\pi^2$ we find

$$\frac{\Delta^{(3)}w_{\mathrm{L}}[u,d]}{w_{\mathrm{L}}[u,d]} = -\frac{(0.18\ \mathrm{GeV})^2}{K^2}.\tag{5.43}$$

5.1.3 Models of the Structure Functions

Although the results for the non-perturbative corrections to the longitudinal and transversal structure functions $w_{\mathrm{L,T}}$ discussed so far, are derived for large values of K^2, we may use them to *model* the behavior of those functions for arbitrary K^2, including $K^2 \sim \Lambda_{\mathrm{QCD}}^2$. To this end, we adopt the logic of large-N_c QCD , discussed in Sect. 3.2, and describe Green's functions by infinite sums of narrow hadronic resonances. Hence, we make an Anzats

$$w_{\mathrm{L,T}} = \sum_i \frac{g_{i,\mathrm{L(T)}}}{K^2 + m_{i,\mathrm{L(T)}}^2},\tag{5.44}$$

where $g_{i,\mathrm{L(T)}}$ and $m_{i,\mathrm{L(T)}}$ are couplings and masses of known hadrons. We choose those hadrons in such a way that their spin, parity and isospin or SU(3)-flavor quantum numbers match those quantum numbers of the functions $w_{\mathrm{L,T}}$. This implies that pseudoscalar mesons contribute to the longitudinal structure function w_{L}, whereas the transversal structure function w_{T} receives contributions from pseudovector mesons. In addition, vector mesons can contribute to both structure functions in the spirit of vector meson dominance. In general, the axial and electromagnetic currents are given by a combination of terms with different isospin and SU(3)-flavor quantum numbers. On the other hand, those quantum numbers are good characteristics of light hadrons. Hence, when constructing hadronic models for the structure functions, it is convenient to decompose the currents into components with definite isospin and/or SU(3) quantum numbers and to deal with those components separately. Finally, any hadronic model has to satisfy the OPE constraints when extrapolated to large values of K^2. Since we derived only a few terms in the OPE in the previous section, we model the structure functions using minimal number of hadrons in (5.44).

Formally expanding (5.44) in power series in $1/K^2$, and comparing the results with the large-K^2 asymptotics of the structure functions, we derive sum rules that the couplings $\{g_i\}$ and the masses of the resonances $\{m_i\}$ should satisfy. For the longitudinal structure function $w_{\mathrm{L}}[u,d]$, we know two terms in the operator product expansion; hence we derive two equations

$$\sum_i g_{i,\mathrm{L}} = 2 \,, \qquad \frac{\sum_i g_{i,\mathrm{L}} m_{i,\mathrm{L}}^2}{\sum_i g_{i,\mathrm{L}}} = (180 \text{ MeV})^2 \,. \qquad (5.45)$$

In case of the first generation, the model for the longitudinal structure function where only the exchange of the pion is retained,

$$w_{\mathrm{L}}[u,d] = \frac{2}{K^2 + m_\pi^2} \,, \qquad (5.46)$$

fits the two sum rules (5.45) reasonably well. Hence, we can use the expression for $w_{\mathrm{L}}[u,d]$ (5.46) as a model for the longitudinal structure function valid for arbitrary K^2. This model smoothly interpolates between perturbative and non-perturbative regimes since it satisfies the constraints from the operator product expansion and is consistent with the expectation that at $K \sim \Lambda_{\mathrm{QCD}}$ the major consequence of the explicit chiral symmetry breaking is a shift of the pion mass to a non-zero value.

In a similar way we construct a *non-perturbative model* for the function $w_{\mathrm{T}}[u,d]$. In this case, we know the large-K^2 expansion through terms $\mathcal{O}(K^{-6})$ and, therefore, can fix three parameters. We write

$$w_{\mathrm{T}} = \sum_i^{\infty} \frac{g_{i,\mathrm{T}}}{K^2 + m_{i,\mathrm{T}}^2} \,. \qquad (5.47)$$

Expanding $w_{\mathrm{T}}(K^2)$ at large K^2, and comparing the subsequent terms in the expansion with the asymptotics (5.39), we derive

$$\sum_i g_{i,\mathrm{T}} = 1, \qquad \sum_i g_{i,\mathrm{T}} m_{i,\mathrm{T}}^2 = (180 \text{ MeV})^2, \qquad \sum_i g_{i,\mathrm{T}} m_{i,\mathrm{T}}^4 = -(710 \text{ MeV})^4. \tag{5.48}$$

Note that we have set $\alpha_s(K) = 1$ in (5.39).

Instead of solving the three equations in (5.48), we approximate the transversal structure function by a linear combination of the propagators that describe exchanges of the ρ and ω vector mesons (neglecting their mass difference) and the $a_1(1260)$ axial vector meson. Note that the isospin quantum numbers of those mesons are identical to the isospin quantum numbers of the electromagnetic and axial currents. We make an Anzats

$$w_{\mathrm{T}}[u,d] = \frac{1}{m_{a_1}^2 - m_\rho^2} \left(\frac{m_{a_1}^2 - m_\pi^2}{Q^2 + m_\rho^2} - \frac{m_\rho^2 - m_\pi^2}{Q^2 + m_{a_1}^2} \right). \tag{5.49}$$

The transversal function w_{T} (5.49) satisfies two first sum rules in (5.48) by construction. Since (5.49) fully determines the model expression for the transversal structure function w_{T}, the $1/K^6$ term in the large-K^2 expansion of w_{T} becomes the prediction of the model that can be compared with the estimate provided by the OPE. Upon expanding (5.49) in $1/K^2$, we find $-(0.96 \text{ GeV})^4$ as the coefficient of the $1/K^6$ term; this should be compared with the OPE estimate $-(0.71 \text{ GeV})^4$, (5.39). Although the agreement is not perfect, we find it reasonable given rather crude estimate of the matrix element of the operator $\mathcal{O}_{\alpha\beta}^6$, (5.33).

Next, we turn to the second generation where the s quark is the only quark which can be considered light. We should be careful with the isospin or, more generally, SU(3) quantum numbers. For the first generation, the weak axial current is $\bar{u}\gamma_\nu\gamma_5 u - \bar{d}\gamma_\nu\gamma_5 d$, so that its quantum numbers coincide with that of the pion. On the contrary, the axial current for the second generation $-\bar{s}\gamma_5\gamma_\nu s$ is a mixture of the SU(3) octet and singlet,

$$- \bar{s}\gamma_\nu\gamma_5 s = \frac{1}{3}\left(\bar{u}\gamma_\nu\gamma_5 u + \bar{d}\gamma_\nu\gamma_5 d - 2\bar{s}\gamma_\nu\gamma_5 s\right) - \frac{1}{3}\left(\bar{u}\gamma_\nu\gamma_5 u + \bar{d}\gamma_\nu\gamma_5 d + \bar{s}\gamma_\nu\gamma_5 s\right), \tag{5.50}$$

Therefore, we write the longitudinal function as

$$w_{\mathrm{L}}[s] = \frac{2}{3}\left(\frac{1}{K^2 + m_\eta^2} - \frac{2}{K^2 + m_{\eta'}^2} \right), \tag{5.51}$$

where the first, octet, term is normalized at large K^2 to $(2N_c/3)(Q_u^2 + Q_d^2 - 2Q_s^2)$, while the second, singlet, term to $(-2N_c/3)(Q_u^2 + Q_d^2 + Q_s^2)$, in correspondence with (5.50). The comparison of the $\mathcal{O}(K^{-4})$ term obtained by expanding (5.51) in powers of $1/K^2$ with the OPE expectation (5.41) shows that the model for the function $w_{\rm L}[s]$ is not very accurate; indeed, the coefficient of the $1/K^4$ term, obtained from expanding (5.51), is $(1\text{ GeV})^2$ while the OPE calculation yields $(0.55\text{ GeV})^2$. This problem may be related to uncalculated terms in the OPE, associated with the mixing of the singlet axial current with two gluons; such terms are important since they are not proportional to the mass of the strange quark m_s.

For the transversal function $w_{\rm T}[s]$ we use the following model

$$w_{\rm T}[s] = -\frac{1}{3}\frac{1}{m_{f_1}^2 - m_\phi^2}\left(\frac{m_{f_1}^2 - m_\eta^2}{K^2 + m_\phi^2} - \frac{m_\phi^2 - m_\eta^2}{K^2 + m_{f_1}^2}\right),\qquad (5.52)$$

where $\phi(1019)$ and $f_1(1426)$ are isoscalar vector and vector-axial mesons relevant for the $s\bar{s}$ channel. Upon expanding (5.52) in inverse powers of K^2, we find that the coefficient of the K^{-6} term evaluates to $(0.79\text{ GeV})^4$ which should be compared with $(0.53\text{ GeV})^4$, predicted by the operator product expansion.

We have constructed the structure functions $w_{\rm L,T}$ for the first and second quark generations that are consistent with constraints from the OPE in QCD and are *non-perturbative* by nature. We can use these structure functions to compute the corrections to the muon anomalous magnetic moment caused by diagrams with anomalous fermion triangles. Our starting point is (5.5) that expresses the correction to the muon magnetic anomaly in terms of the transversal and longitudinal structure functions $w_{\rm L,T}$. Since we have determined those functions, either by explicit perturbative computation or by constructing models that conform with constraints based on the OPE, it is straightforward to compute $a_\mu^{\rm ew,\Delta}$. Such a computation is described in [7] and we refer to that reference for details. Here, we restrict ourselves to logarithmic estimates based on (5.6). We point out that the logarithmic approximation is too crude to provide accurate results for $a_\mu^{\rm ew,\Delta}$; because of that in the final result for $a_\mu^{\rm ew}$ we employ the results of [7]. However, the logarithmic approximation is useful since it illustrates all the physics relevant for the computation of $a_\mu^{\rm ew,\Delta}$ and it is also simple enough so that the details of the calculation can be presented. It is convenient to address the contribution of each generation of quarks and leptons to $a_\mu^{\rm ew,\Delta}$ separately.

We begin with the third generation where all quark masses are large compared to the QCD scale $\Lambda_{\rm QCD}$; because of that, perturbative calculations are reliable. Working in the logarithmic approximation, we write

$$a_\mu^{\rm ew,\Delta}[\tau, t, b] = \left(\frac{\alpha}{\pi}\right)\frac{G_F m^2}{8\pi\sqrt{2}}\left\{\int\limits_{m^2}^{m_t^2}\frac{{\rm d}K^2}{K^2}\left[-2\theta(K^2 - m_\tau^2) - \frac{2}{3}\theta(K^2 - m_b^2)\right]\right.$$

$$\left. -\int\limits_{m^2}^{m_Z^2}\frac{{\rm d}K^2}{K^2}\left[\theta(K^2 - m_\tau^2) + \frac{1}{3}\theta(K^2 - m_b^2)\right] + \frac{2m_Z^2}{9m_t^2}\int\limits_{m_Z^2}^{m_t^2}\frac{{\rm d}K^2}{K^2}\right\}.$$

$$(5.53)$$

The last term in (5.53) is the top quark contribution to the transversal function w_T; the range of logarithmic integration in this case is $m_Z^2 \ll K^2 \ll m_t^2$ and we used (5.8) for the small-K^2 asymptotics of w_T. Performing integrations in (5.53), we arrive at the following contribution of the third generation to the muon anomalous magnetic moment in the logarithmic approximation

$$a_\mu^{\mathrm{ew},\Delta}[\tau, t, b] = -\frac{G_F m^2}{8\pi^2 \sqrt{2}} \left(\frac{\alpha}{\pi}\right) \left[\frac{8}{3} \ln \frac{m_t^2}{m_Z^2} - \frac{2m_Z^2}{9m_t^2} \ln \frac{m_t^2}{m_Z^2} + 4 \ln \frac{m_Z^2}{m_b^2} + 3 \ln \frac{m_b^2}{m_\tau^2}\right].$$

$$(5.54)$$

In case of first and second generations, the fermion masses are small compared to the mass of the Z-boson; hence the factor $m_Z^2/(Q^2 + m_Z^2)$ in (5.6) can be set to one. For the first generation, the longitudinal and transversal structure functions are given by (5.46, 5.49) for up and down quarks and by

$$w_L[e] = 2w_T[e] = -\frac{2}{K^2}\,, \qquad (5.55)$$

for the electron. Note that the electron mass can be neglected there since (5.6) shows that the contribution of (5.55) to the muon magnetic anomaly is determined by $K^2 \gtrsim m^2 \gg m_e^2$. We can also simplify hadronic longitudinal structure function to $w_L[u, d] = 2/K^2$ since $m_\pi^2 \sim m_\mu^2$ and the logarithmic integration in (5.6) is cut off at $K^2 = m^2$. It follows that the sum of the longitudinal functions for the first generation $w_L[e, u, d] = w_L[e] + w_L[u, d]$ *vanishes* and, hence, $w_L[e, u, d]$ does not contribute to $a_\mu^{\mathrm{ew},\Delta}$ in the logarithmic approximation. For the transversal function $w_T[u, d]$ we use (5.49), where we assume $m_{a_1} \gg m_\rho \gg m_\pi$. Then, in the logarithmic approximation, we may write

$$a_\mu^{\mathrm{ew},\Delta}[e, u, d] = \frac{G_F m^2}{8\pi^2 \sqrt{2}} \left(\frac{\alpha}{\pi}\right) \int \frac{dQ^2}{Q^2} \left(-\theta(Q^2 - m^2) + \theta(Q^2 - m_\rho^2)\right)$$

$$= -\frac{G_F m^2}{8\pi^2 \sqrt{2}} \left(\frac{\alpha}{\pi}\right) \int\limits_{m^2}^{m_\rho^2} \frac{dQ^2}{Q^2} = -\frac{G_F m^2}{8\pi^2 \sqrt{2}} \left(\frac{\alpha}{\pi}\right) \ln \frac{m_\rho^2}{m^2}\,. \qquad (5.56)$$

How well does this result compare to the calculations with constituent quarks? In case when the quark masses are taken to be equal to $\bar{m} = 300$ MeV, the contribution of the first generation in the logarithmic approximation reads

$$a_\mu^{\mathrm{ew},\Delta}[e, u, d] = -\frac{G_F m^2}{8\pi^2 \sqrt{2}} \left(\frac{\alpha}{\pi}\right) \ln \left(\frac{\bar{m}^2}{m^2}\right) (2 + 1)\,. \qquad (5.57)$$

The two terms, shown separately in (5.57) refer to the contribution of the longitudinal and transversal functions, respectively. Note, that in contrast to the hadronic computation discussed above, in the constituent quark case the longitudinal structure function gives a *larger* contribution to a_μ than the transversal part; this is the consequence of the introduction of constituent quark masses that fails to properly account for the fact that the pion remains light if the chiral symmetry breaking is introduced properly. Numerically, the difference between (5.56) and (5.57) is at the -1×10^{-11} level.

A similar analysis applies to the second generation. Treating the charm quark as heavy, we cut off its contributions to both w_{L} and w_{T} at $Q^2 \approx 4m_c^2 \approx m_{J/\psi}^2$. The structure functions $w_{\mathrm{L,T}}[s]$ are described above. Integrating over K^2 in the logarithmic approximation, we derive

$$a_\mu^{\mathrm{ew},\Delta}[\mu, c, s] = -\frac{G_F m^2}{8\pi^2 \sqrt{2}} \left(\frac{\alpha}{\pi}\right) \left(4 \ln \frac{m_{J/\psi}^2}{m_\phi^2} + \frac{5}{3} \ln \frac{m_\phi^2}{m_\eta^2} + 3 \ln \frac{m_\eta^2}{m^2}\right). \quad (5.58)$$

We emphasize that the logarithmic estimates presented above are too crude. For example, the logarithmic approximation for the first generation, (5.56) gives $a_\mu[e, u, d] = -1 \times 10^{-11}$, whereas the computation beyond the logarithmic approximation leads to -2×10^{-11}. The calculation of $a_\mu^{\mathrm{ew},\Delta}$ *beyond* the logarithmic approximation is described in [7], where it is shown that the difference between the QCD-based analysis of $a_\mu^{\mathrm{ew},\Delta}$ and the constituent quark model result is, approximately, 2×10^{-11}. Although the study of hadronic effects in electroweak corrections does not lead to large numerical changes compared to the use of constituent quark masses in perturbative calculations, the final result becomes much more credible. In addition, the OPE-based methods for studying hadronic effects discussed in this section will be used in Chap. 6 to analyze a more difficult problem of hadronic light-by-light scattering contribution to the muon magnetic anomaly.

5.2 Final Results for the Electroweak Corrections

In this section we summarize the results for the electroweak corrections to the muon anomalous magnetic moment. We take the numerical value for the electroweak corrections from [4, 7],

$$a_\mu^{\mathrm{ew}} = 154(1)(2) \times 10^{-11} \ . \quad (5.59)$$

The first error in (5.59) corresponds to the hadronic uncertainty and the second to the allowed Higgs mass range, $114 \ \mathrm{GeV} \le m_H \le 250 \ \mathrm{GeV}$, the current uncertainty in the top quark mass and the unknown three-loop electroweak effects.

Finally, we did not mention yet another electroweak correction to the muon magnetic anomaly where perturbative calculations require refinements. Such correction involves $\gamma - Z$ mixing shown in Fig. 5.5. In principle, the $\gamma - Z$ mixing diagram should be studied using the data on e^+e^- annihilation to hadrons in a way that we described in Chap. 3, but the modification, compared to the constituent quark mass estimate in Chap. 4, is negligible

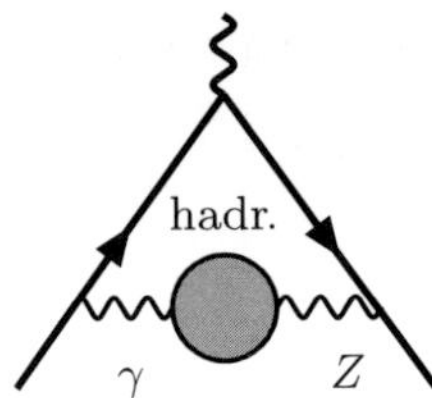

Fig. 5.5. Contribution to the muon anomalous magnetic moment from the $\gamma - Z$ mixing

for the phenomenology of the muon magnetic anomaly. The reason is that the magnitude of the $\gamma - Z$ mixing diagram is strongly suppressed by $g_V^\mu \sim 1 - 4\sin^2\theta_W \sim 0.1$.

References

1. S. L. Adler, Phys. Rev. **177**, 2426 (1969);
   J. S. Bell and R. Jackiw, Nuovo Cimento A **60**, 47 (1969).
2. S. L. Adler and W. A. Bardeen, Phys. Rev. **182**, 1517 (1969).
3. G. 't Hooft, in *Recent Developments In Gauge Theories*, Eds. G. 't Hooft et al., (Plenum Press, New York, 1980).
4. A. Czarnecki, B. Krause and W. J. Marciano, Phys. Rev. Lett. **76**, 3267 (1996).
5. S. Peris, M. Perrottet and E. de Rafael, Phys. Lett. B **355**, 523 (1995).
6. M. Knecht, S. Peris, M. Perrottet and E. de Rafael, JHEP **11**, 003 (2002).
7. A. Czarnecki, W. J. Marciano and A. Vainshtein, Phys. Rev. D **67**, 073006 (2003).
8. S. L. Adler, Phys. Rev. **177**, 2426 (1969);
   L. Rosenberg, Phys. Rev. **129**, 2786 (1963).
9. T. V. Kukhto, E. A. Kuraev, A. Schiller and Z. K. Silagadze, Nucl. Phys. B **371**, 567 (1992).
10. A. Vainshtein, Phys. Lett. B **569**, 187 (2003).
11. B.L. Ioffe and A. Smilga, Nucl. Phys. B **232**, 109 (1984).
12. M. Gell-Man, R. J. Oakes and B. Renner, Phys. Rev. **175**, 2195 (1968).

6 Hadronic Light-by-light Scattering and the Muon Magnetic Anomaly

Among all hadronic contributions to the muon anomalous magnetic moment, the hadronic light-by-light scattering is the one which is most difficult to analyze. This is the consequence of two facts. First, the hadronic light-by-light scattering is largely determined by non-perturbative QCD. Second, it is difficult to relate the hadronic light-by-light scattering contribution to the muon magnetic anomaly to experimental data. Because of these problems, we have to rely on theoretical considerations to estimate the hadronic light-by-light scattering contribution to the muon anomalous magnetic moment.

For the discussion of the hadronic light-by-light scattering, it is important to distinguish between model-dependent and model-independent considerations. Model-dependent considerations are based upon low-energy models that describe interactions of hadrons with photons; quite often, such models are not rooted in QCD. Unfortunately, since the non-perturbative sector of QCD is not well understood, there are very few features of the hadronic light-by-light scattering that can be stated in a model-independent way. Since, in addition to that, it is not easy to control the accuracy of the model-dependent considerations, estimating the central value *and* the theoretical uncertainty of the hadronic light-by-light scattering contribution to a_μ is difficult. It is conceivable that a fully model-independent calculation of the hadronic-light-by-light scattering contribution will never be performed and, hence, the practical issue in connection with $a_\mu^{\rm lbl}$ is to find a way to minimize and control the model-dependence of the theoretical estimates of $a_\mu^{\rm lbl}$.

In spite of these difficulties, current understanding of the hadronic light-by-light scattering contribution to the muon magnetic anomaly is relatively self-consistent. Both low- and high-energy asymptotics of the light-by-light scattering amplitude are understood and taken into account, when models of this amplitude are constructed. It is important that these asymptotics either follow directly from QCD or reflect well-understood symmetries of strong interactions. However, it is clear from the onset, that the approach to the problem based on asymptotics *only* is intrinsically limited; for example, improvements in the precision of such an approach beyond certain point are meaningless.

Calculation of the hadronic light-by-light scattering contribution to the muon anomalous magnetic moment has long and interesting history [1, 2, 3,

K. Melnikov and A. Vainshtein: *Theory of the Muon Anomalous Magnetic Moment*
STMP **216**, 121–144 (2006)
© Springer-Verlag Berlin Heidelberg 2006

4, 5, 6, 7, 8, 9, 10, 11]. Early calculations based on constituent quark approximation were always treated with caution since the effect of strong interactions on the result was unclear. Later, application of effective low-energy theories of strong interactions to the computation of the hadronic light-by-light scattering contribution to a_μ received broad recognition. Because such a description of strong interactions is valid at low energies *only*, somewhere along the way the lore appeared that hadronic light-by-light scattering contribution to a_μ is determined, almost entirely, by low-energy degrees of freedom. Unfortunately, this is incorrect. In the next two sections we give some arguments that demonstrate that the hadronic light-by-light scattering contribution to the muon magnetic anomaly is sensitive to high- and intermediate-momentum regions.[1] We then construct the low-energy model of the hadronic light-by-light scattering [11] that incorporates model-independent constraints that follow from short-distance properties of QCD and approximate chiral symmetry of strong interactions. We use this model to estimate the hadronic light-by-light scattering contribution to a_μ.

6.1 Calculating the Hadronic Light-by-light Scattering Contribution: An Overview

In this section we review, following [11], the calculation of the hadronic light-by-light scattering contribution to the muon anomalous magnetic moment. As we explained in the Introduction to this chapter, such calculations are performed using theoretical models that describe interactions of photons with hadrons at low energies. It is useful to have a theoretical parameter that controls the validity of a model. Since perturbation theory is not an option, we must look for a parameter other than the QCD coupling constant; the two possibilities are the proximity of the chiral symmetry at low energies and the large number of colors N_c [3, 4, 12]. The relevance of those parameters can be seen from the schematic expression for a_μ^{lbl},

$$a_\mu^{\text{lbl}} \sim \left(\frac{\alpha}{\pi}\right)^3 \left[c_1 \frac{m^2}{m_\pi^2} + c_2 N_c \frac{m^2}{\Lambda_{\text{QCD}}^2}\right], \tag{6.1}$$

where it is assumed that $m_\pi > m$. Only the power dependence on m_π^2 is shown; chiral logarithms $\sim \ln m_\pi$ are included into the coefficients $c_{1,2}$. The first, chirally enhanced term is due to loops of charged pions in the light-by-light scattering, Fig. 6.1a. The second, N_c–enhanced, term is due to exchanges of neutral pion or heavier resonances, Fig. 6.1b.

Because the mass of the muon is small, it is natural to expect the chiral parameter $m_\pi^2/(4\pi F_\pi)^2$ to be a better expansion parameter for a_μ^{lbl}. This

---

[1] In this chapter, by "high" and "intermediate" scales we mean energy scales that are larger than or comparable to the mass of the ρ mesons.

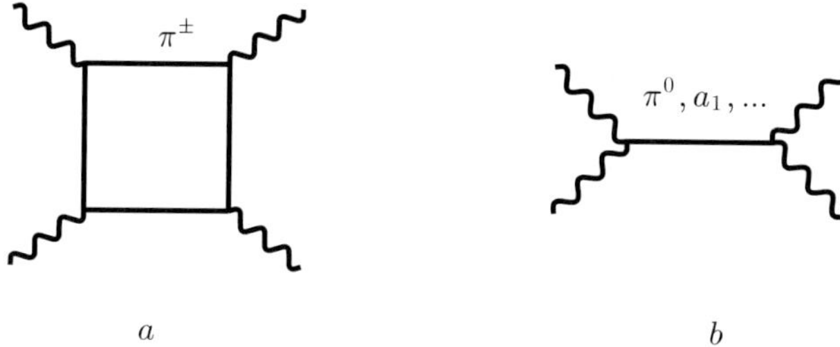

Fig. 6.1. Hadronic contributions to the light-by-light scattering (**a**) charged pion loop, (**b**) exchange of neutral pion and other resonances.

certainly would be valid if the mass of the pion (and the mass of the muon as well) is an order of magnitude smaller while Λ_{QCD} remains the same. In the real world, however, a more careful analysis indicates that things work differently. In what follows, we will show that the chirally enhanced charged pion contribution is always much smaller than the N_c-enhanced contribution; although never proven from first principles, this seems to be a common conclusion to the host of studies of the hadronic light-by-light scattering contribution to the muon magnetic anomaly within various models of low-energy hadronic interactions [3, 4, 5].

Moreover, we observed a similar situation when discussing the hadronic vacuum polarization contribution to a_μ, Chap. 3. There, we have seen that the chirally enhanced two-pion contribution is, approximately, 4×10^{-9} which should be compared with 50×10^{-9}, the N_c-enhanced contribution due to the ρ-meson exchange. Although we do not have a clear understanding of why the chirally enhanced terms are sub-dominant to such an extent, we will take the dominance of the large-N_c expansion over the chiral expansion as the working hypothesis and build our description of the hadronic light-by-light scattering contribution to the muon magnetic anomaly around it.

As we explained in Sect. 3.2.1, the special feature of the large-N_c QCD is that any scattering amplitude can be written as an infinite sum of resonances. This feature helps in constructing a model for hadronic light-by-light scattering but it is insufficient. To constrain the model further, we require the short-distance behavior of the light-by-light scattering amplitude to be consistent with QCD. We derive the corresponding QCD prediction by observing that at large Euclidean photon momenta the operator product expansion (OPE) is applicable to the hadronic light-by-light scattering amplitude. In the following sections, we will show that the leading term in this OPE comes from the quark box diagram enhanced by N_c; hence, the OPE constraints are consistent with the large-N_c limit. Therefore, we require that an acceptable large-N_c hadronic model, extrapolated to large Euclidean photon momenta, matches the perturbative light-by-light scattering amplitude. We find that the *minimal* large-N_c model which satisfies this criterion includes exchanges of the pseudoscalar 0^- mesons π^0, η, η' and the pseudovector 1^+ resonances

a_1, f_1, f_1^* . It is important to emphasize at this point that the model with a finite number of resonances is consistent with the short-distance constraints for a_μ^{lbl}; it is known that this can not be guaranteed for a generic physical observable [13].

The short-distance QCD constraints are most restrictive in the pseudo-scalar isovector channel. This happens for the following reason. The kinematics of the light-by-light scattering relevant for the muon magnetic anomaly necessarily involves soft photon that represents the external magnetic field. Because of that, the four-point function that describes the light-by-light scattering in a most general kinematics, reduces to a three-point function. In a special kinematic limit of this three-point function, when invariant masses of two virtual photons are much larger than the invariant mass of the third one, this channel is fully described by the exchange of the neutral pion. This description is non-perturbative since it works for an arbitrary small invariant mass of the third virtual photon, in spite of the fact that, in general, the OPE applies only when that mass is much larger than Λ_{QCD}.

Before discussing the calculation of the hadronic light-by-light scattering contribution to the muon magnetic anomaly in detail, we present examples that illustrate our claim that a_μ^{lbl} is sensitive to large momenta. We do so in two steps. First, we revisit the calculation of the light-by-light scattering in QED and address it in the context of an effective field theory. After that, we compute a_μ^{lbl} keeping only terms that are enhanced by *both*, the factor N_c and two powers of the chiral logarithm $\ln m_\pi$. While these examples are sufficiently simple to be discussed in detail, they demonstrate interesting subtleties associated with the computation of the hadronic light-by-light scattering contribution to the muon magnetic anomaly.

6.2 The Light-by-light Scattering Contribution to the Muon Anomaly in QED

To understand the role of the high-momentum region in the computation of a_μ^{lbl}, it is useful to revisit QED. In particular, we want to discuss the τ-lepton contribution to the light-by-light scattering component of a_μ, considered in Chap. 2, from a different perspective.

As we discussed in Chap. 2, the QED light-by-light scattering contributions to a_μ are computed directly, starting from relevant Feynman diagrams, Fig. 2.6, and using the mass hierarchy $m_\tau \gg m_\mu$ to simplify the calculation along the way. An alternative way to make use of the mass hierarchy is to "integrate out" the τ lepton from the QED Lagrangian and to employ the resulting *effective field theory* to compute the light-by-light scattering contribution to the muon magnetic anomaly.

In general, to construct an effective field theory [14] we require matching calculations where certain Green's functions are computed in the full theory and in the effective theory; the Lagrangian of the effective field theory is adjusted in such a way that the calculations agree. What makes calculations in the effective field theory advantageous is the fact that one can determine the Lagrangian of the effective theory by doing matching calculations for a limited number of simple Green's functions. Since we are interested in computing the τ-lepton contribution to the light-by-light component of the three-loop QED corrections to a_μ using effective field theory methods, we have to establish what information from the full theory is required. Once this question is answered, it becomes clear which loop momenta contribute to the final result.

Let us assume that for the light-by-light scattering diagrams the leading contribution to $a_\mu^{\rm lbl}$, computed as an expansion in powers of m/m_τ, comes from the momentum configuration where all photon momenta k are small, $k \sim m \ll m_\tau$. Integrating out the τ lepton in this case induces the Euler-Heisenberg Lagrangian

$$\mathcal{L}_{EH} = \frac{\alpha^4}{360 m_\tau^4} \left[4 \left(F_{\mu\nu} F^{\mu\nu} \right)^2 + 7 \left(F_{\mu\nu} \tilde{F}^{\mu\nu} \right)^2 \right] , \qquad (6.2)$$

where $\tilde{F}_{\mu\nu} = (1/2)\epsilon_{\mu\nu\alpha\beta} F^{\alpha\beta}$ is the dual of the electromagnetic field-strength tensor. What is the contribution of $\mathcal{L}_{EH}$ to the light-by-light scattering component of the muon magnetic anomaly? By dimensional arguments, it is easy to see that the range of loop momenta $k \lesssim m$ produces [2]

$$a_\mu^{\rm lbl,\tau} \sim \left(\frac{\alpha}{\pi} \right)^3 \frac{m^4}{m_\tau^4} . \qquad (6.3)$$

Comparing this to the exact result in Sect. 2.5, (2.87),

$$a_\mu^{\rm lbl,\tau} = \left(\frac{\alpha}{\pi} \right)^3 \left\{ \left(\frac{3}{2} \zeta_3 - \frac{19}{16} \right) \frac{m^2}{m_\tau^2} + \mathcal{O}\left(\frac{m^4}{m_\tau^4} \right) \right\} , \qquad (6.4)$$

we find the mismatch in powers of m_τ. This implies that, upon integrating out the τ lepton, we relevant missed operators whose mass dimensions are *lower* than the mass dimension of $\mathcal{L}_{EH}$.

Upon reflection, it is easy to identify the operator that has been missed. Somewhat surprisingly, it is the *muon magnetic dipole operator itself*; its contribution to the effective low-energy Lagrangian can be written as

$$\mathcal{L} = C \frac{m}{m_\tau^2} \bar{\mu} \sigma_{\alpha\beta} \mu F_{\alpha\beta} , \qquad (6.5)$$

where C is the Wilson coefficient. The Wilson coefficient is determined by photon virtualities of order m_τ and therefore can only be computed in the full theory (in our case, QED with both τ and μ leptons). Also, there is no

---

[2] The integration is actually divergent at large k; we use m as the upper cut-off.

observable simpler than the anomalous magnetic moment, from which the Wilson coefficient C could have been determined.

The example we just considered demonstrates that modes with large virtualities are important for the light-by-light scattering component of the muon magnetic anomaly. This conclusion will be important when the *hadronic* light-by-light scattering contribution to a_μ is studied. Since hadrons are strongly interacting particles, the exact computation of the light-by-light scattering amplitude is not possible; the only way to approach the problem is to use chiral perturbation theory at low energies. However, such a description is necessarily insufficient and the effective theory should be extrapolated to larger virtualities to provide the estimate of the Wilson coefficient C. In the next section we describe how this is done *in the logarithmic approximation*. After that, we construct a model of the hadronic light-by-light scattering amplitude that interpolates between small and large photon virtualities and, therefore, enables us to estimate the Wilson coefficient C.

6.3 Hadronic Light-by-light Scattering: Logarithmic Terms

As we will show in this section, the N_c-enhanced contribution to the hadronic light-by-light scattering component of the muon anomalous magnetic moment is additionally enhanced by $\ln(\Lambda/m_\pi)$, where $\Lambda \sim 1$ GeV is a typical hadronic scale [15]. Such an enhancement appears because of the mass gap in QCD between the mass of the pion m_π and the masses of other hadrons ρ, ω, η, ϕ etc. that are of order Λ. For our purposes, it is important that the mass of the muon m is comparable to the mass of the pion and is also much smaller than the scale Λ.

To describe physics at the energy scale $E \sim m_\pi \sim m \ll \Lambda$, we can use an effective field theory obtained by integrating out heavy degrees of freedom in QCD. The result is the chiral effective theory [16], extended to include muons and photons.

As we discussed in Sect. 6.1, there are two parameters that can be used to organize calculations in this effective theory. The first one is a typical energy of the process, the standard expansion parameter in chiral perturbation theory. In addition, there is another parameter, the number of colors $N_c = 3$ which *can be* considered to be large. In this section we show that terms enhanced by N_c play an important role in the hadronic light-by-light scattering component of a_μ. This can be understood qualitatively if one realizes that N_c-enhanced contribution of the neutral pion to the muon magnetic anomaly contains an additional enhancement by chiral logarithms [15],

$$a_\mu^{\text{lbl}} \sim \left(\frac{\alpha}{\pi}\right)^3 \frac{N_c m^2}{\Lambda^2} \ln^2 \frac{\Lambda}{m_\pi} \,. \tag{6.6}$$

It is convenient to approach the calculation of the logarithmically enhanced terms from the chiral effective theory; we need part of this theory that describes the interaction of photons with hadrons. The leading term in the chiral expansion is the Lagrangian of scalar QED

$$\mathcal{L} = -\frac{1}{4}F_{\mu\nu}F^{\mu\nu} + D_\mu\pi^\dagger D^\mu\pi - m_\pi^2|\pi|^2 \ . \tag{6.7}$$

Here $D_\mu = \partial_\mu - ieA_\mu$ is the covariant derivative and π is the charged pion field. If this Lagrangian is used to compute the contribution of the charged pion loop to a_μ, the result is chirally enhanced (i.e. scales as $1/m_\pi^2$) but it is not enhanced by N_c. Also, it does not contain any logarithmic terms, sensitive to the cut-off scale Λ since QED with both fermions and scalars is the renormalizable theory. The chirally enhanced contribution to a_μ^{lbl} is discussed in Sect. 6.7.

Higher terms in the chiral expansion contain the Wess–Zumino–Witten term [17]. For a single pion and two photons this is the dimension-5 operator which describes the interaction of a neutral pion with two photons as a consequence of the Adler–Bell–Jackiw (ABJ) anomaly [20]

$$\mathcal{L}_{ABJ} = \frac{\alpha N_c}{12\pi F_\pi}\, F_{\mu\nu}\tilde{F}^{\mu\nu}\pi^0 \ . \tag{6.8}$$

This is the only N_c-enhanced term in the chiral Lagrangian that contributes to the hadronic light-by-light component of the muon magnetic anomaly at order α^3. It leads to the following $\pi^0\gamma\gamma$ interaction vertex

$$-i\,\frac{\alpha N_c}{3\pi F_\pi}\, \epsilon_{\mu\nu\alpha\beta}q_1^\alpha q_2^\beta \epsilon_1^\mu \epsilon_2^\nu \ , \tag{6.9}$$

where $\epsilon_{1,2}$ are the photon polarization vectors and both photon momenta are taken to be incoming. Using this interaction vertex, it is easy to construct three Feynman diagrams with a virtual π^0 contribution to the hadronic light-by-light scattering component of the muon magnetic anomaly at order $(\alpha/\pi)^3$. Such diagrams, shown in Fig. 6.2, are usually referred to as the "pion-pole" contribution.

By simple power counting, it is easy to see that the planar diagram, Fig. 6.2a, diverges while the non-planar diagram Fig. 6.2b is finite. In addition to an overall divergence in the planar diagram that occurs when both loop momenta in the diagram are large, there is also a divergent subgraph that describes radiatively induced pion-muon coupling. Because of this sub-divergence, the planar pion-pole diagram exhibits a *double logarithmic* sensitivity to the ultra-violet cut-off of the theory Λ.

It is not difficult to compute the $\ln^2(\Lambda/m_\pi)$-enhanced terms in the pion-pole diagram Fig. 6.2a. We begin with the subgraph that describes the one-loop contribution to $\pi^0\mu^+\mu^-$ vertex,

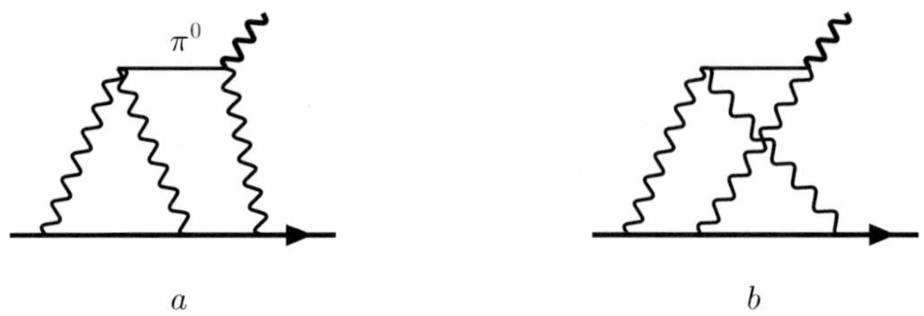

Fig. 6.2. The "pion-pole" contribution to the hadronic light-by-light scattering component of a_μ. The second planar diagram symmetric to the diagram a is not shown

$$\mathcal{M}[\pi^0(k_1) + \mu(p) \to \mu(p + k_1)]$$

$$= -\frac{4\alpha^2 N_c}{3F_\pi} \int^\Lambda \frac{\mathrm{d}^4 k_2}{(2\pi)^4} \frac{\epsilon_{\mu\nu\alpha\beta} k_1^\alpha k_2^\beta}{k_2^2(k_1 - k_2)^2} \bar{u}_{p+k_1} \gamma^\mu \frac{1}{\hat{p} + \hat{k}_2 - m} \gamma^\nu u_p \ . \tag{6.10}$$

The integration over k_2 is simple in the logarithmic approximation,

$$\int^\Lambda \frac{\mathrm{d}^4 k_2}{(2\pi)^4} \frac{k_2^\beta}{k_2^2(k_2 - k_1)^2} \frac{1}{\hat{p} + \hat{k}_2 - m} \approx \frac{i}{64\pi^2} \gamma^\beta \ln\left(\frac{\Lambda^2}{-k_1^2}\right) \ . \tag{6.11}$$

Using $\epsilon_{\alpha\mu\beta\nu}\gamma^\mu\gamma^\beta\gamma^\nu = 3!\, i\gamma^\alpha\gamma_5$ and $\bar{u}_{p+k_1}\hat{k}_1\gamma_5 u_p = 2m\, u_{p+k_1}\gamma_5 u_p$ we arrive at

$$\mathcal{M}[\pi^0(k_1) + \mu(p) \to \mu(p + k_1)] = -\left(\frac{\alpha}{\pi}\right)^2 \frac{N_c}{4F_\pi} m\, \bar{u}_{p+k_1}\gamma_5 u_p \ln\left(\frac{\Lambda^2}{-k_1^2}\right) \ . \tag{6.12}$$

We use this result to get the amplitude for the diagram Fig. 6.2a by integrating over k_1,

$$\mathcal{M}^{\mathrm{pl}} = -ie\left(\frac{\alpha}{\pi}\right)^3 \frac{N_c^2 m}{6F_\pi^2} \int^\Lambda \frac{\mathrm{d}^4 k_1}{(2\pi)^4} \frac{\tilde{f}_{\nu\beta} k_1^\beta}{k_1^2(k_1^2 - m_\pi^2)} \ln\left(\frac{\Lambda^2}{-k_1^2}\right) \bar{u}_p \gamma^\nu \frac{1}{\hat{p} + \hat{k}_1 - m} \gamma_5 u_p$$

$$= -ie\left(\frac{\alpha}{\pi}\right)^3 \frac{N_c^2 m}{192\pi^2 F_\pi^2} \ln^2\frac{\Lambda}{m_\pi} \tilde{f}_{\nu\beta} \bar{u}_p \sigma^{\nu\beta}\gamma_5 u_p \ . \tag{6.13}$$

Here we retain only terms linear in the momentum of the external photon q which is present implicitly in the dual electromagnetic field-strength $\tilde{f}_{\nu\beta} = \epsilon_{\nu\beta\alpha\mu} q^\alpha \epsilon^\mu$.

Using the identity $\bar{u}_p \sigma^{\alpha\beta}\gamma_5 u_p \tilde{f}_{\alpha\beta} = 2i\bar{u}_p \sigma^{\alpha\beta} u_p \epsilon_\alpha q_\beta$, we arrive at the following "pion-pole" contribution to the muon anomalous magnetic moment [6, 7]

$$a_\mu^{\text{lbl}}[\pi^0] = \left(\frac{\alpha}{\pi}\right)^3 3\left(\frac{N_c}{3}\right)^2 \frac{m^2}{(4\pi F_\pi)^2} \ln^2\frac{\Lambda}{m_\pi} + \dots , \qquad (6.14)$$

where the ellipses stands for uncalculated terms that contain *at most* single logarithms of the ultra-violet cut-off, $\mathcal{O}(\ln \Lambda/m)$. Note that since $F_\pi \sim N_c^{1/2}$, the pion pole contribution to the muon magnetic anomaly scales like $a_\mu^{\text{lbl}}[\pi^0] \sim N_c$. It is appropriate to make a few comments at this point.

(1) The double-logarithmic approximation for the pion-pole contribution to a_μ, (6.14), is the unique prediction of QCD that follows from its chiral properties.

(2) In the context of the low-energy effective field theory, the apparent divergence $\ln^2(\Lambda)$ in (6.14) is removed by adding counter-terms to the effective Lagrangian

$$\mathcal{L}_{\text{ct}} = C_1 im\,\bar{\mu}\gamma_5\mu\,\pi^0 + C_2 m\,\bar{\mu}\sigma_{\alpha\beta}\mu F_{\alpha\beta} , \qquad (6.15)$$

where $C_1 \propto \ln(\Lambda/\mu)$, $C_2 \propto \ln^2(\Lambda/\mu)$ and the mass scale μ is associated with hadron dynamics at around 1 GeV. A peculiar feature of these counter-terms is that they contain the magnetic dipole operator itself; this implies that the muon magnetic anomaly can not be computed from the low-energy effective field theory *alone*. A similar situation arises in the τ-lepton light-by-light scattering contribution to a_μ considered in the previous section. Because it is not possible to extract C_2 from any observable that is simpler than the muon anomalous magnetic moment itself, it has to be estimated from full QCD; this fact makes the theory of the hadronic light-by-light scattering contribution to a_μ highly non-trivial.

(3) Physically, the scale Λ is identified with the mass of the ρ meson or heavier resonances. Taking $\Lambda = m_\rho = 770$ MeV in (6.14), we obtain $a_\mu^{\text{lbl}}[\pi^0] = 120 \times 10^{-11}$. However, existing calculations [8, 10] of the hadronic light-by-light scattering contribution seem to indicate that logarithmically enhanced terms provide too crude an estimate of the hadronic light-by-light scattering contribution to the muon magnetic anomaly.

(4) Equation (6.14) has been used to check the numerical evaluation of the hadronic light-by-light scattering contribution to a_μ [6, 7] and thus contributed to uncovering the sign error in earlier calculations of the hadronic light-by-light component of the muon anomalous magnetic moment.

Finally, we mention that it is possible to extend the calculation of $a_\mu^{\text{lbl}}[\pi^0]$ in such a way that terms enhanced by a single logarithm of an ultra-violet cut-off are included. To achieve that, we need to extend the calculation described in this section to compute the induced $\pi^0\mu^+\mu^-$ coupling, (6.11), *up to a constant term*, in the limit of large k_1. This can be done only in the model-dependent way. Once the constant term is fixed, the subsequent integration over k_1 gives both, the $\mathcal{O}(\ln^2 \Lambda)$ and $\mathcal{O}(\ln \Lambda)$ contributions to $a_\mu^{\text{lbl}}[\pi^0]$. Such a calculation is described in [10].

6.4 Model for the Hadronic Light-by-light Scattering

In this section, we construct the model for the hadronic light-by-light scattering that we use to compute the muon anomalous magnetic moment. We begin with the description of the kinematics of the light-by-light scattering relevant for the muon magnetic anomaly. The light-by-light scattering amplitude involves four photons with momenta q_i and the polarization vectors ϵ_i. We take the photon momenta to be incoming, $\sum q_i = 0$. The first three photons are virtual, while the fourth photon represents the external magnetic field and can be regarded as the real photon with the vanishingly small momentum q_4. The amplitude $\mathcal{M}$ is defined as

$$\mathcal{M} = \alpha^2 N_c \, \mathrm{Tr}\left[\hat{Q}^4\right] \mathcal{A} = \alpha^2 N_c \, \mathrm{Tr}\left[\hat{Q}^4\right] \mathcal{A}_{\mu_1\mu_2\mu_3\gamma\delta} \epsilon_1^{\mu_1} \epsilon_2^{\mu_2} \epsilon_3^{\mu_3} f^{\gamma\delta} \tag{6.16}$$

$$= -e^3 \int \mathrm{d}^4x \, \mathrm{d}^4y \, e^{-iq_1 x - iq_2 y} \, \epsilon_1^{\mu_1} \epsilon_2^{\mu_2} \epsilon_3^{\mu_3} \langle 0|T\left\{ j_{\mu_1}(x) j_{\mu_2}(y) j_{\mu_3}(0) \right\} |\gamma\rangle \, ,$$

where j_μ is the hadronic electromagnetic current, $j_\mu = \bar{q}\hat{Q}\gamma_\mu q$, written in terms of the three quark flavors $q = \{u, d, s\}$ with $\hat{Q}$ being the 3×3 diagonal matrix of quark electric charges, $\hat{Q} = \mathrm{diag}(2/3, -1/3, -1/3)$. As usual, $f^{\gamma\delta} = q_4^\gamma \epsilon_4^\delta - q_4^\delta \epsilon_4^\gamma$ denotes the field-strength tensor of the soft photon; the light-by-light scattering amplitude is proportional to this tensor due to gauge invariance. To compute the muon anomalous magnetic moment, we need the amplitude $\mathcal{M}$ through first order in q_4; since $f_{\alpha\beta} \sim q_4$, we can set $q_4 = 0$ in the tensor amplitude $\mathcal{A}_{\mu_1\mu_2\mu_3\gamma\delta}$ and calculate it assuming that $q_1 + q_2 + q_3 = 0$ for the virtual photons. This equality implies that the momenta q_1, q_2, q_3 form a triangle; because of that there are just three independent Lorentz invariant variables that we choose to be the virtualities of the photons q_{1-3}^2.

In general, the light-by-light scattering amplitude is a complicated function of photon virtualities. However, there are only *two* distinct kinematic regimes in the light-by-light scattering amplitude. The first (symmetric) one occurs when the Euclidean momenta of the three photons are comparable in magnitude $q_1^2 \sim q_2^2 \sim q_3^2$; the second, asymmetric, happens when one of the photon momenta is much smaller than the other two. The asymmetric limit can be analyzed in a simple way using the operator product expansion. In addition, this limit is important because it helps to identify the pole-like structures in the OPE amplitudes and in this way connect the OPE to phenomenological models of the light-by-light scattering. Schematically,

$$\mathcal{A}_{\mathrm{OPE}} \sim \frac{c}{q^2} \to \frac{c}{q^2 - M^2} \, , \tag{6.17}$$

where M is the mass of the meson with appropriate quantum numbers.

6.4.1 Operator Product Expansion and Triangle Amplitude

Since the light-by-light scattering amplitude is symmetric with respect to photon permutations, we can study the asymmetric kinematic configuration assuming that $q_1^2 \approx q_2^2 \gg q_3^2$. In this regime, we begin with the well-known OPE (see e.g. [18]) for the product of two electromagnetic currents that carry the largest momenta q_1, q_2,

$$\int \mathrm{d}^4x\, \mathrm{d}^4y\, \mathrm{e}^{-iq_1 x - iq_2 y}\, T\left\{j_\mu(x) j_\nu(y)\right\} = \int \mathrm{d}^4z\, \mathrm{e}^{iq_3 z}\, \frac{2\epsilon_{\mu\nu\delta\rho}\, \hat{q}^\delta}{\hat{q}^2}\, j_5^\rho(z) + \cdots .$$

(6.18)

Here, $j_5^\rho = \bar{q}\hat{Q}^2 \gamma^\rho \gamma_5\, q$ is the axial current, where different flavors enter with weights proportional to squares of their electric charges, $q_3 = -q_1 - q_2$ and $\hat{q} = (q_1 - q_2)/2 \approx q_1 \approx -q_2$. We retain only the leading (in the limit of large Euclidean $\hat{q}$) term in the OPE associated with the axial current j_5^ρ; the ellipsis in (6.18) stands for sub-leading terms suppressed by powers of $\Lambda_{\mathrm{QCD}}/\hat{q}$. The momentum $q_1 + q_2 = -q_3$ flowing through j_ρ^5 is assumed to be much smaller than $\hat{q}$. As a side remark, note that (6.18) has been applied earlier in various situations; for example, the matrix element of (6.18) between the pion and the vacuum states gives the asymptotic behavior of the $\pi^0 \gamma^* \gamma^*$ amplitude at large photon virtualities [19].

In what follows, we use matrix elements of (6.18) to connect the hadronic light-by-light scattering amplitude with the QCD prediction. For this reason, it is convenient to write (6.18) in a way that facilitates such connection. To this end, we express the axial current in (6.18) as a linear combination of axial currents with definite SU(3) quantum numbers. In particular, we introduce the isovector, $j_{5\rho}^{(3)} = \bar{q}\lambda_3 \gamma_\rho \gamma_5 q$, hypercharge, $j_{5\rho}^{(8)} = \bar{q}\lambda_8 \gamma_\rho \gamma_5 q$, and the SU(3) singlet, $j_{5\rho}^{(0)} = \bar{q}\gamma_\rho \gamma_5 q$, currents, and write

$$j_{5\rho} = \sum_{a=3,8,0} \frac{\mathrm{Tr}\left[\lambda_a \hat{Q}^2\right]}{\mathrm{Tr}\left[\lambda_a^2\right]}\, j_{5\rho}^{(a)} ,$$

(6.19)

where λ_0 is the unity matrix

Since the dependence of the product of the two vector currents on the largest momenta $q_{1,2}$ is factored out in (6.18), the next step is to find the dependence of the light-by-light scattering amplitude on the momentum q_3. This dependence is given by the amplitudes $T_{\gamma\rho}^{(a)}$ that involve axial currents $j_{5\rho}^{(a)}$ and two electromagnetic currents, one with momentum q_3 and the other one (the external magnetic field) with the vanishing momentum

$$T_{\mu3\rho}^{(a)} = i\, \langle 0|\int \mathrm{d}^4z\, \mathrm{e}^{iq_3 z} T\{j_{5\rho}^{(a)}(z)\, j_{\mu3}(0)\}|\gamma\rangle .$$

(6.20)

We have discussed the triangle amplitudes for such kinematics in Chap. 5 and much of what has been said there applies to the discussion in this section. In Chap. 5 we have shown that $T_{\mu3\rho}^{(a)}$ can be written in terms of two independent amplitudes, $w_{\mathrm{L}}^{(a)}$ and $w_{\mathrm{T}}^{(a)}$,

$$T^{(a)}_{\mu 3 \rho} = -\frac{ie\, N_c \mathrm{Tr}\,[\lambda_a \hat{Q}^2]}{4\pi^2} \left\{ w^{(a)}_{\mathrm{L}}(q_3^2)\, q_{3\rho} q_3^\sigma \tilde{f}_{\sigma\mu 3} \right.$$

$$\left. + w^{(a)}_{\mathrm{T}}(q_3^2)\left(-q_3^2 \tilde{f}_{\mu 3 \rho} + q_{3\mu 3} q_3^\sigma \tilde{f}_{\sigma\rho} - q_{3\rho} q_3^\sigma \tilde{f}_{\sigma\mu 3}\right) \right\}. \tag{6.21}$$

The first (second) amplitude is related to the longitudinal (transversal) part of the axial current, respectively. In terms of hadrons, the invariant function $w_{\mathrm{L(T)}}$ describes the exchanges of pseudoscalar (pseudovector) mesons.

In perturbation theory, $w_{\mathrm{L,T}}$ are computed from triangle diagrams with two vector currents and an axial current. For massless quarks, we obtain

$$w^{(a)}_{\mathrm{L}}(q^2) = 2w^{(a)}_{\mathrm{T}}(q^2) = -\frac{2}{q^2}. \tag{6.22}$$

An appearance of the longitudinal part is the signature of the axial Adler-Bell-Jackiw anomaly [20]. Although perturbation theory is only reliable for $q^2 \gg \Lambda^2_{\mathrm{QCD}}$, where it coincides with the leading term of the OPE for the time-ordered product of the axial and electromagnetic currents, expressions for the longitudinal functions $w^{(3,8)}_{\mathrm{L}}$ given in (6.22) are *exact* QCD results in the chiral limit $m_q = 0$ for non-singlet axial currents. Because both perturbative [21] and nonperturbative [22] corrections to the axial anomaly are absent, (6.22) is correct all the way down to small q^2, where poles in q^2 are associated with Goldstone pseudoscalar mesons, π^0 in $w^{(3)}_{\mathrm{L}}$ and η in $w^{(8)}_{\mathrm{L}}$.

The absence of perturbative and non-perturbative corrections and therefore the possibility to use the OPE expressions for vanishing values of q^2 is unique for the longitudinal part of the non-singlet axial current. For the transversal functions w_{T} as well as for the singlet longitudinal function $w^{(0)}_{\mathrm{L}}$, there are higher order terms in the OPE that, upon summation, generate mass terms that shift the pole position away from $q^2 = 0$, $1/q^2 \to 1/(q^2 - M^2)$; we discussed in detail how this happens in Chap. 5. We use such modifications of the pole-like terms for each channel in what follows. The lightest pseudovector mesons are the $a_1(1260)$, $f_1(1290)$ and $f_1^*(1420)$ mesons. For the singlet axial current, the pole in $w^{(0)}_{\mathrm{L}}$ is shifted to $m^2_{\eta'}$.

Consider a triangle amplitude for any isospin channel in the limit of large q^2, where the OPE and the perturbation theory are applicable and (6.22) is valid. An important consequence of this equation is that triangle amplitudes are *not* suppressed at large q^2. In terms of hadrons it means that no form factor is present in the $h\gamma^*\gamma$ interaction vertex where the real photon is soft (external magnetic field). We note that this conclusion contradicts the common practice [3, 4, 6], rooted in the application of low-energy models to the hadronic light-by-light scattering amplitude when, for the π^0 exchange, the form factor $F_{\pi\gamma^*\gamma}(q^2, 0)$ is introduced. Such transition form factor has to be present when one of the photons is virtual, the other photon is on the mass shell and the pion is on the mass shell as well. However, this is not the kinematics that corresponds to the fermion triangle diagram in the light-by-light scattering amplitudes relevant for a_μ computation, where the pion virtuality is *the same* as the virtuality of one of the photons.

Combining (6.18–6.21), we write the light-by-light scattering amplitude $\mathcal{A}_{\mu_1\mu_2\mu_3\gamma\delta}$ for $q_1^2 \approx q_2^2 \gg q_3^2$ in the following form

$$\mathcal{A}_{\mu_1\mu_2\mu_3\gamma\delta}f^{\gamma\delta} = \frac{8}{\hat{q}^2}\epsilon_{\mu_1\mu_2\delta\rho}\hat{q}^\delta \sum_{a=3,8,0} W^{(a)} \left\{ w_{\mathrm{L}}^{(a)}(q_3^2)\, q_3^\rho q_3^\sigma \tilde{f}_{\sigma\mu_3} \right.$$

$$\left. +w_{\mathrm{T}}^{(a)}(q_3^2)\left(-q_3^2\tilde{f}_{\mu_3}^\rho + q_{3\mu_3}q_3^\sigma \tilde{f}_\sigma^\rho - q_3^\rho q_3^\sigma \tilde{f}_{\sigma\mu_3}\right) \right\} + \cdots , \tag{6.23}$$

where no hierarchy between q_3^2 and Λ_{QCD}^2 is assumed. The weights $W^{(a)}$ refer to different SU(3) quantum numbers; they read

$$W^{(a)} = \frac{\left(\mathrm{Tr}\,[\lambda_a \hat{Q}^2]\right)^2}{\mathrm{Tr}\,[\lambda_a^2]\mathrm{Tr}\,[\hat{Q}^4]}\,; \qquad W^{(3)} = \frac{1}{4}\,, \quad W^{(8)} = \frac{1}{12}\,, \quad W^{(0)} = \frac{2}{3}\,.$$

In the limit $q_3^2 \gg \Lambda_{\mathrm{QCD}}^2$, (6.23) can be simplified using the asymptotic expressions (6.22) for the invariant functions $w_{\mathrm{L,T}}^{(a)}$. Contracting the tensor amplitude with the photon polarization vectors and analytically continuing to Euclidean space, we arrive at

$$\mathcal{A} = \frac{4}{q_3^2\,\hat{q}^2}\{f_2\tilde{f}_1\}\{\tilde{f}f_3\} - \frac{4}{q_3^2\,\hat{q}^4}\left(\{q_2 f_2\tilde{f}_1\tilde{f}f_3 q_3\} + \frac{q_1^2+q_2^2}{4}\{f_2\tilde{f}_1\}\{\tilde{f}f_3\}\right) + \cdots .$$

$$\tag{6.24}$$

Here, $f_i^{\mu\nu} = q_i^\mu \epsilon_i^\nu - q_i^\nu \epsilon_i^\mu$ are the field-strength tensors; braces denote either traces of products of the matrices $f_i^{\mu\nu}$ or their contractions with vectors q_i.

Starting from (6.24), we use Euclidean notations instead of Minkowskian ones. The continuation to Euclidean space mostly concerns the change in sign for all q_i^2 and the overall change in sign for the amplitude $\mathcal{A}$, since it involves the product of two Levi-Cevita tensors. We note that for arbitrary q_i^2, the light-by-light scattering amplitude can be written in terms of nineteen independent tensor structures and five independent form-factors; the corresponding expressions can be found in [11].

6.4.2 The Model

Different terms in (6.23) can be identified with exchanges of the pseudoscalar (pseudovector) mesons described by the functions $w_{\mathrm{L(T)}}^{(a)}(q_3^2)$. Extrapolating (6.24) from $q_{1,2}^2 \gg \Lambda_{\mathrm{QCD}}^2$ to arbitrary $q_{1,2}^2$, we arrive at the following model

$$\mathcal{A} = \mathcal{A}_{\mathrm{PS}} + \mathcal{A}_{\mathrm{PV}} + \text{permutations} , \tag{6.25}$$

where

$$\mathcal{A}_{\mathrm{PS}} = \sum_{a=3,8,0} W^{(a)} \phi_{\mathrm{L}}^{(a)}(q_1^2, q_2^2) w_{\mathrm{L}}^{(a)}(q_3^2) \{f_2 \tilde{f}_1\}\{\tilde{f} f_3\}, \tag{6.26}$$

$$\mathcal{A}_{\mathrm{PV}} = \sum_{a=3,8,0} W^{(a)} \phi_{\mathrm{T}}^{(a)}(q_1^2, q_2^2) w_{\mathrm{T}}^{(a)}(q_3^2) \left(\{q_2 f_2 \tilde{f}_1 \tilde{f} f_3 q_3\} + \{q_1 f_1 \tilde{f}_2 \tilde{f} f_3 q_3\} \right.$$
$$\left. + \frac{q_1^2 + q_2^2}{4} \{f_2 \tilde{f}_1\}\{\tilde{f} f_3\} \right). \tag{6.27}$$

The form factors $\phi_{\mathrm{L,T}}^{(a)}(q_1^2, q_2^2)$ account for the dependence of the amplitude on $q_{1,2}^2$. Pictorially, Fig. 6.1b, these form factors can be associated with the interaction vertex for the two virtual photons on the left hand side of the diagram, whereas the meson propagator and the interaction vertex on the right hand side form the triangle amplitude described by the functions $w_{\mathrm{L,T}}^{(a)}(q_3^2)$. In the following sections we introduce models for these functions consistent with the short-distance behavior of the light-by-light scattering amplitude.

Note that our model does not include explicit exchanges of vector or scalar mesons. This is a consequence of the fact that, to leading order, the OPE of two vector currents produces an axial vector current only. However, the vector mesons are present in our model implicitly, through the momentum dependence of the form factors $\phi_{L,T}^{(a)}$ as well as the transversal functions $w_{\mathrm{T}}^{(a)}$.

6.5 Constraints on the Pseudoscalar Exchange

The π^0 exchange provides the largest fraction of the hadronic light-by-light scattering contribution to a_μ. It is therefore important to scrutinize this component as much as possible and ensure that it satisfies all the constraints that follow from first principles.

As we discussed earlier, the longitudinal part of the triangle amplitude is fixed by the ABJ anomaly. Accounting for explicit violation of the chiral symmetry given by the small mass of the pion, we derive

$$w_{\mathrm{L}}^{(3)}(q^2) = \frac{2}{q^2 + m_\pi^2} . \tag{6.28}$$

The ABJ anomaly also fixes the form factor $\phi_{\mathrm{L}}^{(3)}(0,0)$,

$$\phi_{\mathrm{L}}^{(3)}(0,0) = \frac{N_c}{4\pi^2 F_\pi^2} , \tag{6.29}$$

so that the model for the pion exchange in the light-by-light scattering amplitude takes the form,

$$\mathcal{A}_{\pi^0} = -\frac{N_c W^{(3)}}{2\pi^2 F_\pi^2} \frac{F_{\pi\gamma^*\gamma^*}(q_1^2, q_2^2)}{q_3^2 + m_\pi^2} \{f_2 \tilde{f}_1\}\{\tilde{f} f_3\} + \text{permutations} . \tag{6.30}$$

In spite of the fact that we refer to (6.30) as the "pion pole" contribution, it describes the *complete*, on- and off-shell, light-by-light scattering amplitude in the pseudoscalar isotriplet channel.

The $\pi\gamma^*\gamma^*$ form factor $F_{\pi\gamma^*\gamma^*}(q_1^2, q_2^2)$ is defined as

$$F_{\pi\gamma^*\gamma^*}(q_1^2, q_2^2) = \frac{\phi_{\rm L}^{(3)}(q_1^2, q_2^2)}{\phi_{\rm L}^{(3)}(0,0)} \, . \tag{6.31}$$

The comparison with the OPE constraint given by the relevant term in (6.23) leads to

$$\lim_{q^2 \gg \Lambda_{\rm QCD}^2} F_{\pi\gamma^*\gamma^*}(q^2, q^2) = \frac{8\pi^2 F_\pi^2}{N_c \, q^2} \, . \tag{6.32}$$

Equation (6.32) is the correct asymptotics of the pion transition form factor [19], as can be easily seen by taking the matrix element of the correlator of two electromagnetic currents (6.18) between the vacuum and the single pion state and using (6.35) to compute the matrix element of the axial isovector current. We therefore conclude that the neutral pion exchange in (6.26) saturates the corresponding short-distance QCD constraint.

This comparison also proves our previous claim that the form factor $F_{\pi\gamma^*\gamma}(q_3^2, 0)$ cannot be present in the amplitude (6.30); if that form factor is introduced, the asymptotics of the light-by-light scattering amplitude becomes $1/q_3^4$, in conflict with the $1/q_3^2$ behavior that follows from perturbative QCD. This proof is, of course, equivalent to our discussion of the triangle amplitude in Sect. 6.4.

As we mentioned earlier, the absence of the second form factor in the amplitude (6.30) distinguishes our approach from all other calculations of the pion-pole contribution to a_μ that exist in the literature. We show below that this feature has a non-negligible impact on the final numerical result for the pseudoscalar contribution to the muon magnetic anomaly; the pion-pole contribution *increases*, because the absence of the second form factor leads to a slower convergence of the integral over loop momenta, making the result larger.

Further constraints on the model follow from restrictions on the pion transition form factor $F_{\pi\gamma^*\gamma^*}$. The primary role here is played by the CLEO measurement [23] of the reaction $e^+e^- \to e^+e^-\pi^0$ for q^2 in the interval from 1 to 9 GeV^2. Writing the pion form factor as

$$F_{\pi\gamma^*\gamma^*}(q_1^2, q_2^2) = \frac{4\pi^2 F_\pi^2}{N_c} \frac{N(q_1, q_2)}{(q_1^2 + M_1^2)(q_1^2 + M_2^2)(q_2^2 + M_1^2)(q_2^2 + M_2^2)} \, , \tag{6.33}$$

with

$$N(q_1, q_2) = q_1^2 q_2^2 (q_1^2 + q_2^2) - h_2 q_1^2 q_2^2 + h_5(q_1^2 + q_2^2) + N_c M_1^4 M_2^4 / 4\pi^2 F_\pi^2 \, ,$$

and fitting it to the CLEO data, Knecht and Nyffeler found [24] $M_1 = 769$ MeV, $M_2 = 1465$ MeV, $h_5 = 6.93$ GeV^4. The parameter h_2 was not determined in [24] because the dominant contribution to CLEO data comes from the kinematic configuration where one of the two photons that collide to produce a neutral pion is quasi-real, $q_{1,2}^2 = 0$. However, it is possible to determine h_2 using the following consideration.

Consider (6.33) in the symmetric limit $q_1^2 = q_2^2 \gg \Lambda_{\rm QCD}^2$. It is easy to see that the non-zero h_2 contributes to the $1/q^4$ correction to the leading asymptotics of the pion form factor, (6.32). Such correction comes from the twist 4 operators in the OPE of two electromagnetic currents (6.18). It was analyzed long ago in [19] using the OPE and the QCD sum rules approaches. The result of that analysis implies that the coefficient of the $\mathcal{O}(q^{-4})$ term in the asymptotics of the pion form factor is numerically small; in terms of the parametrization (6.33), this means that $h_2 \approx -10 \text{ GeV}^2$ has to be chosen.

Equations (6.30) and (6.33) completely specify the model for the pion pole contribution that we use for numerical calculations below. Before going into that, we mention that the numerical results seem to be fairly stable against small modifications of the model, as long as small- and large-q^2 asymptotics are not violated. A more detailed discussion of this issue can be found in [11].

We turn to numerical consequences of the accurate matching between low- and high-energy parts of the light-by-light scattering amplitude. A convenient technique for numerical integration of the pion-pole light-by-light scattering amplitude is described in [6]. Using the model for the pion-pole contribution (6.30) with $h_2 = -10 \text{ GeV}^2$, we find $a_\mu^{\rm lbl}[\pi^0] = 76.5 \times 10^{-11}$. Compared to the central value $a_\mu^{\rm lbl}[\pi^0] = 58 \times 10^{-11}$, derived in [6], we observe an increase by, approximately, 20×10^{-11}.

A similar analysis for the isosinglet channels leads to the conclusion that these channels are saturated by η and η' mesons; matching to perturbative QCD asymptotics suggests that no transition form factor is present for the soft photon interaction vertex in those cases as well. Since the isosinglet contributions are smaller than that of π^0, we do not use sophisticated models for η and η' transition form factors and estimate them using the simplest possible vector meson dominance (VMD) form factor.[3] The $\eta(\eta')\gamma^*\gamma^*$ interaction vertex is normalized in such a way that the decay widths of these mesons into two photons is correctly reproduced; this allows to account for the $\eta - \eta'$ mixing in a simple way. We find $a_\mu^{\rm lbl}[\eta] = a_\mu^{\rm lbl}[\eta'] = 18 \times 10^{-11}$. The sum of the contributions from all pseudoscalar mesons (π^0, η, η') leads to the result

$$a_\mu^{\rm lbl}[\pi^0, \eta, \eta'] = 114(10) \times 10^{-11} \ . \tag{6.34}$$

The uncertainty in (6.34) is estimated as follows. We can check the accuracy of our model by comparing its predictions for the two-photon couplings of the η and η' mesons to experimental results. Because of the $\eta - \eta'$ mixing, we expect that the sum of $\eta(\eta')\gamma\gamma$ couplings squared is predicted by

---

[3] The VMD form factors are inconsistent with QCD expectations when virtualities of both photons become large; this, however, is largely irrelevant for final numerical results for η and η' contributions to the muon magnetic anomaly.

the model more accurately than each of the couplings separately. To derive the two-photon couplings to the pseudoscalar, we use (6.21) and (6.22). For example, in case of the π^0 meson, we consider the isovector part of the triangle amplitude $T_{\gamma\rho}^{(3)}$. The residue at $q^2 = 0$, corresponding to π^0 pole, is the product of two matrix elements,

$$\langle 0|j_{5\rho}^{(3)}|\pi^0\rangle = 2iF_\pi\, q_\rho\,, \quad \langle\pi^0|j_{\mu 3}|\gamma\rangle = -4eg_{\pi\gamma\gamma}q^\sigma\tilde{f}_{\sigma\mu 3}\,. \tag{6.35}$$

Comparing with (6.21), (6.22) we derive the well-known result [20] for the $\pi\gamma\gamma$ coupling

$$g_{\pi\gamma\gamma} = \frac{N_c\,\mathrm{Tr}\,[\lambda_3\hat{Q}^2]}{16\pi^2\,F_\pi}\,. \tag{6.36}$$

In a similar way, the $g_{\eta(\eta')\gamma\gamma}$ coupling in the chiral limit can be derived. We find

$$r = \frac{g_{\eta\gamma\gamma}^2 + g_{\eta'\gamma\gamma}^2}{g_{\pi\gamma\gamma}^2} = 3\,, \tag{6.37}$$

whereas using experimental values for the $\eta(\eta')\gamma\gamma$ couplings we arrive at $r = 2.5(1)$. We use this 20% discrepancy as an error estimate on the $\eta + \eta'$ contribution to (6.34); this accounts for eighty percent of the whole uncertainty. However, we note that the comparison between theory and experiment seems to imply that further *increase* in $a_\mu^{\mathrm{lbl}}[\pi^0, \eta, \eta']$ is likely since the agreement between "experimental" and theoretical asymptotics can be improved by adding more pseudoscalar mesons to the model.

The central value in (6.34) is almost 40 percent larger than most of the existing results for a_μ^{lbl} [3, 4, 6]. The major effect comes from removing the form factor from the interaction vertex of the *soft* photon (magnetic field) with the pseudoscalar meson; the necessity to do that *unambiguously* follows from matching the pseudoscalar pole amplitude to the perturbative QCD expression for the light-by-light scattering. On the contrary, the error estimate in (6.34) is subjective; it is based on the variation of the result when input parameters of the model are modified. Nevertheless, we believe that the error estimate in (6.34) adequately reflects the current knowledge of the pseudoscalar contribution to the muon anomalous magnetic moment.

6.6 Pseudovector Exchange

In this section we discuss the pseudovector contribution to the muon anomalous magnetic moment given by the amplitude $\mathcal{A}_{\mathrm{PV}}$, (6.27). From (6.22), (6.23), (6.24), we find the asymptotics of $\phi_{\mathrm{T}}^{(a)}$ and $w_{\mathrm{T}}^{(a)}$,

$$\lim_{q^2\gg\Lambda_{\mathrm{QCD}}^2}\phi_{\mathrm{T}}^{(a)}(q^2,q^2) = -\frac{4}{q^4}\,, \qquad \lim_{q^2\gg\Lambda_{\mathrm{QCD}}^2}w_{\mathrm{T}}^{(a)}(q^2) = \frac{1}{q^2}\,. \tag{6.38}$$

As we mentioned earlier, the lightest pseudovector resonances are the a_1 meson with the mass $M_{a_1} = 1260$ MeV, the f_1 meson, with the mass $M_{f_1} = 1285$ MeV and the f_1^* meson with the mass $M_{f_1^*} = 1420$ MeV. The contribution of these mesons to a_μ^{lbl} is cut off at the scales defined by their masses. This suggests that the pole-like singularities in (6.24) are shifted from zero to the masses of the corresponding pseudovector and vector mesons.

The simplest estimate of the pseudovector contribution can be obtained by assigning equal masses to all pseudovector mesons irrespective of their SU(3) quantum numbers. This leads to the following model for the form factors and the structure functions, consistent with perturbative QCD constraints (6.38),

$$\phi_{\mathrm{T}}^{(a)}(q_1^2, q_2^2) = -\frac{4}{(q_1^2 + M^2)(q_2^2 + M^2)} \,, \qquad w_{\mathrm{T}}^{(a)}(q) = \frac{1}{q^2 + M^2} \,. \qquad (6.39)$$

Although this model is unrealistic, we use it to derive a simple analytic result that exhibits the dependence of the pseudovector contribution on the mass scale M. Assuming that $M \gg m$, we obtain

$$a_\mu^{\mathrm{PV}} = \left(\frac{\alpha}{\pi}\right)^3 \frac{m^2}{M^2} N_c \mathrm{Tr}\,[\hat{Q}^4] \left[\frac{71}{192} + \frac{81 S_2}{16} - \frac{7\pi^2}{144}\right] \approx 1010 \, \frac{m^2}{M^2} \times 10^{-11} \,, \qquad (6.40)$$

where $S_2 = 0.26043$. Using $M = 1300$ MeV as an example, we obtain $a_\mu^{\mathrm{PV}} = 7 \times 10^{-11}$.

Two comments about this result are appropriate. First, we compare it to the existing estimates of the pseudovector meson contribution [3, 4]. In those references, the results 2.5×10^{-11} and 1.7×10^{-11} have been obtained. The difference between our result (6.40) and the results of [3, 4] can be explained by the absence of the form factor for the $\gamma^* \gamma h$ interaction vertex in our model; when such a form factor is introduced, our result decreases to 2.6×10^{-11}, in good agreement with the estimates in [3, 4].

Also, we note that the result (6.40) exhibits strong sensitivity to the mass of the pseudovector meson and the mass parameter in the form factor. If we associate the mass scale M in (6.40) with the mass of the ρ-meson, the result increases roughly by a factor 4 and becomes $a_\mu^{\mathrm{PV}} \sim 28 \times 10^{-2}$. Because of the strong sensitivity to the mass parameter, we have to introduce a more sophisticated model accounting for the mass differences in different SU(3) channels.

We start with the isovector function $w_{\mathrm{T}}^{(3)}$. This function describes the triangle amplitude that involves the isovector axial current, the virtual photon and the soft photon. We expect therefore that $w_{\mathrm{T}}^{(3)}(q_3^2)$ contains two poles with respect to q_3^2; the first one, that corresponds to the $a_1(1260)$ pseudovector meson and the second one, that corresponds to the vector mesons ρ, ω, ϕ, thereby reflecting properties of the virtual photon. Such a model was constructed in [25] where it was required that, for large values of q^2, the equality $w_{\mathrm{L}}(q^2) = 2w_{\mathrm{T}}(q^2)$, remains valid through $\mathcal{O}(q^{-4})$ terms. This requirement leads to

$$w_{\mathrm{T}}^{((3))}(q_3^2) = \frac{1}{m_{a_1}^2 - m_\rho^2} \left[\frac{m_{a_1}^2 - m_\pi^2}{q^2 + m_\rho^2} - \frac{m_\rho^2 - m_\pi^2}{q^2 + m_{a_1}^2} \right], \tag{6.41}$$

where we do not distinguish between the masses of ρ and ω mesons. Correspondingly, the form factor $\phi_{\mathrm{T}}^{((3))}(q_1^2, q_2^2)$ becomes

$$\phi_{\mathrm{T}}^{(3)}(q_1^2, q_2^2) = -\frac{4}{(q_1^2 + m_\rho^2)(q_2^2 + m_\rho^2)}. \tag{6.42}$$

For the isoscalar pseudovector mesons $f_1(1285)$ and $f_1(1420)$ we assume the "ideal" mixing similar to ω and ϕ; this assumption is consistent with experimental data for decays of these resonances. Then, instead of the hypercharge and the SU(3)-singlet weights $W^{(8)}$ and $W^{(0)}$, we use

$$W^{(u+d)} = \frac{25}{36}, \qquad W^{(s)} = \frac{1}{18}, \tag{6.43}$$

and the following expressions for the corresponding functions w_{T} and ϕ_{T}

$$\begin{aligned}
w_{\mathrm{T}}^{(u+d)}(q^2) &= \frac{1}{m_{f_1}^2 - m_\omega^2} \left[\frac{m_{f_1}^2 - (m_\eta^2/5)}{q^2 + m_\omega^2} - \frac{m_\omega^2 - (m_\eta^2/5)}{q^2 + m_{f_1}^2} \right], \\
\phi_{\mathrm{T}}^{(u+d)}(q_1^2, q_2^2) &= -\frac{4}{(q_1^2 + m_\omega^2)(q_2^2 + m_\omega^2)}, \\
w_{\mathrm{T}}^{(s)}(q^2) &= \frac{1}{m_{f_1^*}^2 - m_\phi^2} \left[\frac{m_{f_1^*}^2 + m_\eta^2}{q^2 + m_\phi^2} - \frac{m_\phi^2 + m_\eta^2}{q^2 + m_{f_1^*}^2} \right], \\
\phi_{\mathrm{T}}^{(s)}(q_1^2, q_2^2) &= -\frac{4}{(q_1^2 + m_\phi^2)(q_2^2 + m_\phi^2)}.
\end{aligned} \tag{6.44}$$

Note, that these refinements of the simple expression for the function w_{T} in (6.39) make the effective mass of the pseudovector meson smaller. This leads to the increase in a_μ^{PV} as compared to (6.40). We obtain the following estimate

$$a_\mu^{\mathrm{lbl}}[a_1, f, f^*] = (5.7 + 15.6 + 0.8) \times 10^{-11} = 22 \times 10^{-11}, \tag{6.45}$$

where the three terms displayed separately are due to the isovector, $u + d$ and s exchanges respectively.

To check the stability of the model, we consider an opposite case for the mixing, assuming that f_1 is a pure octet and f_1^* is an SU(3) singlet meson. The estimate for a_μ^{PV} becomes

$$a_\mu^{\mathrm{lbl}}[a_1, f, f^*] = (5.7 + 1.9 + 9.7) \times 10^{-11} = 17 \times 10^{-11}. \tag{6.46}$$

We see that the SU(3)-singlet contribution is significant, in spite of the fact that the corresponding masses are the largest. The reason for such a behavior is a stronger coupling of the SU(3)-singlet meson to two photons. Also, in spite of a very strong redistribution between the different SU(3) channels, the final result for the pseudovector contribution is relatively stable against such variations of the model. We use the result for the pseudovector contribution in (6.45) in our final estimate of $a_\mu^{\rm lbl}$ assigning $\pm 5 \times 10^{-11}$ as the uncertainty estimate.

6.7 The Pion Box Contribution

In this section we discuss another part of $a_\mu^{\rm lbl}$ frequently considered in the literature, the pion box contribution, Fig. 6.1a. This component of the hadronic light-by-light scattering amplitude is peculiar because, being independent of the number of colors N_c, it is enhanced by another large parameter of low-energy QCD, $4\pi F_\pi/m_\pi \sim 10$.

The results for the pion box contribution to $a_\mu^{\rm lbl}$ were obtained in [3, 4]; they are $a_\mu^{\rm lbl}[\pi^\pm] = -4.5(8.5) \times 10^{-11}$ [3] and $a_\mu^{\rm lbl}[\pi^\pm] = -19(5) \times 10^{-11}$ [4]. The difference between the two results is attributed to a different treatment of subleading terms in the chiral expansion; while the extended Nambu–Jona-Lasinio (ENJL) model is used in [4] to couple photons to pions, the so-called hidden local symmetry (HLS) model is used in [3].[4] Although the smallness of $a_\mu^{\rm lbl}[\pi^\pm]$ shows that the chiral enhancement is not efficient for $a_\mu^{\rm lbl}$, strong sensitivity of the final result to a *particular method* of including heavier resonances suggests that the chiral expansion per se may not be a reliable tool to analyze the pion box contribution. If this is true, it is natural to ask to what extent the above estimates of the pion box contribution can be trusted. To answer this question, we investigate the pion box contribution based on the analytic calculation of $a_\mu^{\rm lbl}[\pi^\pm]$ in the framework of the HLS model.

The use of the chiral expansion to estimate subleading $\mathcal{O}(N_c^0)$ contributions to $a_\mu^{\rm lbl}$ is motivated by the following considerations. If the pion box contribution to a_μ is determined by small values of virtual momenta, comparable to the masses of muon and pion, we can compute it by using chiral perturbation theory. The leading term in the chiral expansion delivers a parametrically enhanced contribution $(\alpha/\pi)^3 (m/m_\pi)^2$ to $a_\mu^{\rm lbl}$, which can be computed using the scalar QED Lagrangian for pions (6.7). The Lagrangian (6.7) is the leading term in the effective chiral Lagrangian and, hence, terms neglected in (6.7) are suppressed by the square of the ratio of the pion mass to the scale of the chiral symmetry breaking. Numerically, these corrections

---

[4] The claim in [26] and [3] that the standard VMD violates the Ward identities for the $\gamma^*\gamma^*\pi\pi$ amplitude is not correct, if the VMD is implemented in the standard way, by introducing the factor $(M^2 g_{\mu\nu} + q_\mu q_\nu)/(M^2 + q^2)$ for each photon in any interaction vertex. The Ward identities, discussed in [3], are then automatically satisfied.

are expected to be small since $m_\pi^2/m_\rho^2 \sim 0.04$ and $m_\pi^2/(4\pi F_\pi)^2 \sim 0.025$; therefore, they should not change the scalar QED prediction by more than a few per cent.

However, existing computations of the pion box contribution [3, 4] do not support these expectations. On the contrary, they indicate that the scalar QED contribution is reduced by a factor ranging from three [4] to ten [3], when subleading terms in the chiral expansion are included. One is then inclined to conclude that the chiral expansion for the pion box contribution does not work. In order to identify the reason for that, we computed several terms of the expansion in m_π/m_ρ in the framework of the HLS model. Comparing the magnitude of the subsequent terms in the expansion, we can determine how fast the chiral expansion converges and estimate a typical scale of the loop momentum in the pion box diagram.

As we demonstrate below, typical loop momenta in the pion box diagram are approximately $4m_\pi \sim 500$ MeV $\lesssim m_\rho$; this leads to a slow convergence of the chiral expansion and explains, to a certain extent, a very strong cancelation between the leading order scalar QED result and the first m_π^2/m_ρ^2 correction. The remaining terms in the chiral expansion are smaller, although not negligible.

We point out that large values of typical loop momenta make the use of both, the scalar QED model (6.7) and its modifications based on the VMD, unphysical because quality of such models deteriorates rapidly once the loop momenta exceed the ρ-meson mass. To see that the model fails relatively early, we consider the deep inelastic scattering of a virtual photon with large value of q^2, on a pion. The Lagrangian (6.7) then implies the dominance of the longitudinal structure function, while QCD predicts the opposite. Modifying the scalar QED Lagrangian (6.7) to accommodate the VMD either directly or through the HLS model, does not fix this problem, since only an overall factor $(M_\rho^2/(M_\rho^2 + q^2))^2$ is introduced in the imaginary part of the forward scattering amplitude. This mismatch implies that models used to compute the pion box contribution become unreliable once the energy scale of the order of the ρ-meson mass is passed. Since $4m_\pi$ is only marginally smaller than m_ρ, it is hard to tell how big a mistake we make by ignoring the fact that the hadronic model has an incorrect asymptotic behavior.

The above considerations suggest that while it is most likely that the pion box contribution to a_μ is relatively small, as follows from a strong cancelation of the two first terms in the chiral expansion, the precise value of this contribution is impossible to obtain, using simple VMD and the like models.

The analytic calculation of the pion box contribution within the HLS model [3] was performed in [11] and we refer to that reference for details. Here, we quote the result of that analysis. The pion box contribution to the muon anomalous magnetic moment is written as

$$a_\mu^{\mathrm{lbl}}[\pi^\pm] = \left(\frac{\alpha}{\pi}\right)^3 \sum_{i=0}^{\infty} f_i(m, m_\pi) \left(\frac{m_\pi^2}{m_\rho^2}\right)^i . \tag{6.47}$$

The functions f_i for $i = 0, ..., 4$ can be found in [11]. For numerical estimates we use $m_\pi = 136.98$ MeV, $m = 105.66$ MeV and $m_\rho = 769$ MeV. With these input values, (6.47) evaluates to

$$a_\mu^{\mathrm{lbl}}[\pi^\pm] = -0.0058 \left(\frac{\alpha}{\pi}\right)^3 = (-46.37 + 35.46 + 10.98 - 4.70 - 0.3 + ...) \times 10^{-11}$$

$$= -4.9(3) \times 10^{-11} \, , \tag{6.48}$$

where subsequent terms in (6.48) correspond to subsequent terms in (6.47).

The feature of the result (6.48) which has to be noticed is the strong cancelation between the leading and the two subleading terms in the chiral expansion (6.48); the cancelation is so strong that the final result is almost entirely determined by the fourth, $\mathcal{O}(m_\pi^6/m_\rho^6)$, term in the series. It is the strong cancelation between the leading terms in the chiral expansion that ensures the smallness of the final value of the pion pole contribution to a_μ.

We can use (6.48) to determine typical loop momenta in the pion box contribution. For simplicity, we assume $m_\pi = m$ and obtain

$$a_\mu^{\mathrm{lbl}}[\pi^\pm, m_\pi = m] \approx (-69 + 54 + 18 - 8 - 1 + ..) \times 10^{-11} \, . \tag{6.49}$$

We also assume that this contribution to a_μ can be described by the chiral expansion with the effective scale μ. This scale characterizes the typical virtual momentum in the pion box diagram. Motivated by chiral perturbation theory, we make an Ansatz

$$a_\mu^{\mathrm{lbl}}[\pi^\pm, m_\pi = m] \approx \left(\frac{\alpha}{\pi}\right)^3 \frac{m_\pi^2}{\mu^2} \left(c_1 + c_2 \frac{\mu^2}{m_\rho^2} + c_3 \frac{\mu^4}{m_\rho^4} + ...\right) . \tag{6.50}$$

We further assume that all the coefficients in the above series are numbers of order one. Setting $c_1 = 1$ in the above equation, we determine the value of μ by comparing it with the first term in (6.49). We obtain $\mu = 4.25 m_\pi$. Then, (6.50) becomes

$$a_\mu^{\mathrm{lbl}}[\pi^\pm, m_\pi = m] \approx (-69 + 41c_2 + 24c_3 + 14c_4 + ...) \times 10^{-11} \, , \tag{6.51}$$

which implies that with $c_2 \approx 1.3$, $c_3 \approx 0.75$ and $c_4 \approx -0.6$, we can fit (6.49).

The above calculation suggests a simple way to understand the magnitude of the the chirally suppressed terms in (6.48). Since $\mu \approx 4 m_\pi \approx 550$ MeV $\lesssim m_\rho$, the chiral expansion converges, but rather slowly. Therefore, calculations based on the chiral expansion make sense *in principle*, but are of little help in practice because the results are very sensitive to higher order power corrections. Since none of the models, be it the HLS model or the VMD model, can claim full control over higher order power corrections in the chiral expansion, the *exact result* for the pion box contribution is not very meaningful. However, the fact that the chiral expansion is applicable suggests that the strong cancelation between the leading and *the first* subleading terms in the chiral expansion of $a_\mu^{\mathrm{lbl}}[\pi^\pm]$ may be a generic, model-independent feature.

Therefore, we find it reasonable to *believe* that the pion box contribution to a_μ^{lbl} is much smaller than the estimate based on the chirally enhanced scalar QED result for the pions. However, once this point of view is accepted, the chiral enhancement looses its power as the theoretical parameter and the pion box contribution becomes just one of many $\mathcal{O}(N_c^0)$ contributions about which nothing is known at present. Therefore, for the final estimate of a_μ^{lbl} we use

$$a_\mu^{\mathrm{lbl},\mathrm{N_c^0}} = 0(10) \times 10^{-11} , \qquad (6.52)$$

where the error estimate is very subjective.

6.8 The Hadronic Light-by-light Scattering Contribution to a_μ

In this section we summarize the results for the hadronic light-by-light scattering contribution to the muon magnetic anomaly. We have presented the minimal large-N_c model for the hadronic light-by-light scattering amplitude that incorporates both chiral and short-distance QCD constraints and provides a suitable foundation for numerical estimate of a_μ^{lbl}. We emphasize, however, that even with all the constraints included, the calculation of a_μ^{lbl} is a guesstimate and, hence, some degree of subjectivity is involved in deriving the answer. This, however, is inevitable when strong interactions are involved. We take [11]

$$a_\mu^{\mathrm{lbl}} = 136(25) \times 10^{-11} , \qquad (6.53)$$

as an estimate of the hadronic light-by-light scattering contribution to the muon anomalous magnetic moment. We use this number in the next chapter, where we derive the theoretical prediction for the muon magnetic anomaly in the Standard Model. Before that, we make a few comments on the result in (6.53).

We point out that the result in (6.53) is approximately 50 percent larger, than the value $\sim 80(30) \times 10^{-11}$ obtained in [6, 3, 4] and about 25 percent larger than $a_\mu^{\mathrm{lbl}} = 108 \times 10^{-11}$, obtained in [9]. The error estimate in (6.53) includes the sum of $\mathcal{O}(N_c^0)$ error estimate in (6.52) as well as 15×10^{-11} as an error estimate for the sum of the pseudoscalar and the pseudovector exchanges. The shift in the central value of a_μ^{lbl}, compared to other results in the literature, is a consequence of a better matching of the low-energy hadronic models and the short-distance QCD achieved within the model described in the main body of this chapter.

Finally, we note that (6.53) can be checked for consistency in the following way. We can estimate the light-by-light scattering contribution as a sum of two terms – the pion-pole contribution, to account for low-momentum region and the massive quark box contribution, to account for large-momentum regime. If the quark masses are chosen to be $m_q = 300$ MeV, the result for the quark box contribution is $\sim 60 \times 10^{-11}$. Combining this with the pion pole contribution, we get the estimate $a_\mu^{\text{lbl}} \approx 120 \times 10^{-11}$. Although the above consideration is hardly a proof, it does indicate the tendency of the result for a_μ^{lbl} to increase once the contribution of the large-momentum region is accounted for in the correct way.

References

1. T. Kinoshita, B. Nizic, Y. Okamoto, Phys. Rev. D **31**, 2108 (1985).
2. J. Calmet, S. Narison, M. Perrottet and E. de Rafael, Phys. Lett. B **16**, 283 (1976); Rev. Mod. Phys. **49**, 21 (1977).
3. M. Hayakawa, T. Kinoshita and A. I. Sanda, Phys. Rev. D **54**, 3137 (1996).
4. J. Bijnens, E. Pallante and J. Prades, Nucl. Phys. B **474**, 379 (1996), Nucl. Phys. B **626**, 410 (2002).
5. M. Hayakawa and T. Kinoshita, Phys. Rev. D **57**, 465 (1998); Erratum-ibid. **66**, 019902 (2002).
6. M. Knecht and A. Nyffeler, Phys. Rev. D **65**, 073034 (2002).
7. M. Knecht, A. Nyffeler, M. Perrottet and E. de Rafael, Phys. Rev. Lett. **88**, 071802 (2002).
8. I. Blokland, A. Czarnecki and K. Melnikov, Phys. Rev. Lett. **88**, 071803 (2002).
9. E. Bartos, A. Z. Dubnickova, S. Dubnicka, E. A. Kuraev and E. Zemlyanaya, Nucl. Phys. B **632**, 330 (2002).
10. M. Ramsey-Musolf and M. B. Wise, Phys. Rev. Lett. **89**, 041601 (2002).
11. K. Melnikov and A. Vainshtein, Phys. Rev. D **70**, 113006 (2004).
12. E. de Rafael, Phys. Lett. B **322**, 239 (1994).
13. For a recent discussion, see J. Bijnens, E. Gamiz, E. Lipartia and J. Prades, JHEP **0304**, 055 (2003).
14. H. Georgi, Ann. Rev. Nucl. Part. Phys. Sci., **43**, 209 (1993); I. Z. Rothstein, *TASI lectures on effective field theories*, hep-ph/0308266.
15. K. Melnikov, Int. Journ. Mod. Phys. A **16**, 4591 (2001).
16. S. Weinberg, Physica A **96**, 327 (1979); J. Gasser and H. Leutwyler, Ann. Phys. **158**, 142 (1984); J. Gasser and H. Leutwyler, Nucl. Phys. B **250**, 465 (1985).
17. J. Wess and B. Zumino, Phys. Lett. **37**, 95 (1975); E. Witten, Nucl. Phys. B **223**, 422 (1983).
18. J. D. Bjorken, Phys. Rev. D **1**, 1376 (1970).
19. V. A. Novikov, M. A. Shifman, A. I. Vainshtein, M. B. Voloshin and V.I Zakharov, Nucl. Phys. B **237**, 525 (1984).
20. S. L. Adler, Phys. Rev. **177**, 2426 (1969); J. S. Bell and R. Jackiw, Nuovo Cimento A **60**, 47 (1969).
21. S. L. Adler and W. A. Bardeen, Phys. Rev. **182**, 1517 (1969).
22. G. 't Hooft, in *Recent Developments In Gauge Theories*, Eds. G. 't Hooft et al., (Plenum Press, New York, 1980).
23. J. Gronberg et al. [CLEO collaboration], Phys. Rev. D **57**, 33 (1998).
24. M. Knecht and A. Nyffeler, Eur. Phys. J. C **21**, 659 (2001).
25. A. Czarnecki, W. J. Marciano and A. Vainshtein, Phys. Rev. D **67**, 073006 (2003).
26. M. B. Einhorn, Phys. Rev. D **49**, 1668 (1994).

7 Standard Model Value for the Muon Magnetic Anomaly

In this section we derive the Standard Model prediction for the muon anomalous magnetic moment by combining results given in various places in the text. We write the anomalous magnetic moment as

$$a_\mu^{\text{th}} = a_\mu^{\text{QED}} + a_\mu^{\text{EW}} + a_\mu^{\text{hvp}} + a_\mu^{\text{hvp,NLO}} + a_\mu^{\text{lbl}} \,, \tag{7.1}$$

and use

$$a_\mu^{\text{QED}} = 116\ 584\ 720(1) \times 10^{-11} \,,$$

$$a_\mu^{\text{EW}} = 154(1)(2) \times 10^{-11} \,,$$

$$a_\mu^{\text{hvp}} = 6934(63) \times 10^{-11}, \quad a_\mu^{\text{hvp,NLO}} = -98(1) \times 10^{-11} \,,$$

$$a_\mu^{\text{lbl}} = 136(25) \times 10^{-11} \,. \tag{7.2}$$

The references to the original papers from where these contributions are extracted can be found in the corresponding chapters. Evaluating the sum in (7.1), we obtain

$$a_\mu^{\text{th}} = 116\ 591\ 846(25)(63) \times 10^{-11} \,, \tag{7.3}$$

where the first error refers to the hadronic light-by-light scattering contribution to a_μ and the second combines all other errors in quadratures.

The results of the experimental measurements of the muon anomalous magnetic moment are summarized in Table 7.1. The two most recent BNL results [9, 10] dominate the world average

$$a_\mu^{\text{exp}} = 116\ 592\ 082(55) \times 10^{-11} \,. \tag{7.4}$$

The difference between the experimental result for a_μ (7.4) and the corresponding theory value (7.3) is

$$a_\mu^{\text{exp}} - a_\mu^{\text{th}} = (237 \pm 55_{\text{exp}} \pm 25_{\text{lbl}} \pm 63_{\text{th}}) \times 10^{-11} \,. \tag{7.5}$$

If the errors in (7.5) are combined in quadratures, the difference between the experiment and the theory amounts to 2.7 standard deviations.

At the moment, there is no consensus on whether this 2.7σ discrepancy is a signal of new physics or an indication that the Standard Model theory of

K. Melnikov and A. Vainshtein: *Theory of the Muon Anomalous Magnetic Moment*
STMP **216**, 145–149 (2006)
© Springer-Verlag Berlin Heidelberg 2006

Table 7.1. Experimental results for the muon magnetic anomaly

CERN cyclotron	1961	[1]	μ^+	0.001145(22)
CERN cyclotron	1962	[2]	μ^+	0.001165(5)
1st muon storage ring at CERN	1966	[3]	μ^-	0.001165(3)
1st muon storage ring at CERN	1968	[4, 5]	μ^+	0.00116616(31)
2nd muon storage ring at CERN	1975-77	[6]	$\mu^\pm$	0.001165923(8)
BNL, 1998 data	2000	[7]	μ^+	0.0011659191(59)
BNL, 1999 data	2001	[8]	μ^+	0.0011659202(15)
BNL, 2000 data	2002	[9]	μ^+	0.00116592039(84)
BNL, 2001 data	2004	[10]	μ^-	0.00116592143(83)
World average	2004	[10]	$\mu^\pm$	0.00116592082(55)

the muon magnetic anomaly requires further development. We tend to believe that the Standard Model theory has to be scrutinized further before resorting to new physics explanations. In this regard, further improvements in understanding the hadronic light-by-light scattering are necessary; in particular, this concerns a careful assessment of its uncertainty.

On the experimental side it is very desirable that small discrepancies in the e^+e^- data from CMD-2, SND and KLOE are settled. A new measurement of the e^+e^- hadronic annihilation cross-section in the energy range 1.4 GeV − 2 GeV is required to understand the difference between inclusive and exclusive measurements. Hopefully, such a measurement can be done with the radiative return method at B-factories. The resolution of the puzzling discrepancy between the τ data and the e^+e^- data is probably not crucial for the physics of the muon magnetic anomaly given multiple confirmations of the low-energy e^+e^- data. The resolution of this puzzle is, however, important for better understanding the isospin symmetry violations in low-energy hadron physics.

The largest uncertainty in the theoretical prediction for the muon anomalous magnetic moment is related to the hadronic vacuum polarization. In this regard, it is interesting to ask if other precision observables are sensitive to the same non-perturbative input and, hence, if they can be used to cross-check the values for the hadronic vacuum polarization employed for a_μ evaluation. Two such observables were recently discussed in the literature [11, 12].

Marciano points out [11] that *larger* values of a_μ^{hvp}, required for the agreement with a_μ^{exp}, imply *lower* values of the Higgs boson mass, within the context of the precision electroweak physics. To understand the connection, note that for computing electroweak corrections within the Standard Model, we require three input parameters, in addition to quark masses and the mass of the Higgs boson. A convenient choice is the electromagnetic coupling constant α, the Fermi constant G_F and the mass of the Z-boson M_Z since all of these parameters are measured with the highest precision. Once the values for these parameters are provided by experiment, other physical quantities become theoretically computable. Comparing theoretical predictions with experimental measurements, we may constrain the unknown Standard Model parameters, in particular the Higgs boson mass.

The hadronic vacuum polarization influences the precision electroweak observables through its contribution to the running of the electromagnetic coupling constant, from the low scale where conventional α is defined, to M_Z which is the appropriate scale for electroweak measurements. We define

$$\alpha(M_Z) = \frac{\alpha}{1 - \Delta\alpha_{\text{lept}} - \Delta\alpha_{\text{hadr}}} , \tag{7.6}$$

where

$$\Delta\alpha_{\text{lept, hadr}} = \frac{\alpha}{3\pi} \int ds \, \frac{R_{\text{lept, hadr}}}{s(M_Z^2 - s)} . \tag{7.7}$$

Values of $\Delta\alpha_{\text{lept,hadr}}$ are related to leptonic and hadronic vacuum polarization contributions to the running of the coupling constant. In what follows, we only discuss the hadronic vacuum polarization since the leptonic vacuum polarization is known precisely. The magnitude of $\Delta\alpha_{\text{hadr}}$ can be estimated from both the e^+e^- and the τ data. When appropriate isospin corrections are applied, the τ data leads to a larger value of $\Delta\alpha_{\text{hadr}}$

$$\Delta\alpha_{\text{hadr}} = \begin{cases} 0.02767(16), & e^+e^- ; \\ 0.02782(16), & \tau . \end{cases} \tag{7.8}$$

The difference between the two values for $\Delta\alpha_{\text{hadr}}$ is less than one standard deviation, so the evidence of the disagreement is not very impressive,[1] but it leads to interesting consequences. Consider the prediction for $\sin^2\theta_W$ within the Standard Model

$$\sin^2\theta_W(M_Z)_{\overline{\text{MS}}} = 0.23101 - 2.77 \cdot 10^{-3} \left[\left(\frac{m_t}{178\,\text{GeV}} \right)^2 - 1 \right]$$

$$+ 9.69 \cdot 10^{-3} \left(\frac{\Delta\alpha_{\text{hadr}}}{0.02767} - 1 \right) + 4.908 \cdot 10^{-4} \ln \frac{m_H}{100\,\text{GeV}}$$

$$+ 3.43 \cdot 10^{-5} \ln^2 \frac{m_H}{100\,\text{GeV}} . \tag{7.9}$$

---

[1] The difference between the e^+e^- and the τ data leads to a stronger disagreement in a_μ than in $\alpha(M_Z)$ because the latter is less sensitive to the e^+e^- hadronic annihilation cross-section at low energies.

It follows from (7.9) that for a given value of $\sin^2\theta_W$ and the mass of the top quark m_t, a larger value of $\Delta\alpha_{\text{hadr}}$ implies a smaller value for the Higgs boson mass m_H. Marciano finds [11] $m_H = 71^{+48}_{-32}$ GeV, by using the $\Delta\alpha_{\text{hadr}}$ determined from the e^+e- data and $m_H = 64^{+44}_{-30}$ GeV, by using the τ data. Although both values of the Higgs boson mass are not in contradiction with the LEP bound $m_H \geq 114$ GeV, a correlation between a better agreement of a_μ^{th} with the experimental value and a *smaller* value of the Higgs boson mass is interesting.

Another observation is due to Maltman [12]. It is known that data on τ decays provides one of the most precise determinations of the strong coupling constant $\alpha_s(M_Z)$ through the theoretical prediction for $\Gamma(\tau \to \nu_\tau + \text{hadrons})$ computed within the framework of the perturbative QCD and the operator product expansion. It gives $\alpha_s(M_Z) = 0.120(2)$. On the other hand, one can perform the same analysis [12] using the spectral density from $e^+e^- \to$ hadrons; this leads to the result $\alpha_s(M_Z) = 0.115(2)$. The difference between the two values for α_s is yet another reflection of the difference between the e^+e^- and the τ data.

The two values of α_s can be compared to other determinations of the strong coupling constant, unrelated to the e^+e^- and τ data. Comparing the above determinations of $\alpha_s(M_Z)$ with the world average[2] taken to be $0.120(2)$, Maltman concludes that the e^+e^- data is disfavored. Note, however, that this conclusion depends on the exact value of α_s that is assumed for the world average which consists of many measurements with different status of systematic errors. For example, if we select α_s measurements that have total error less than 8×10^{-3} and for which NNLO description in perturbative QCD is available, we obtain $\alpha_s(M_Z) = 0.1180(25)$. Such value of α_s lies between the value of α_s obtained from the e^+e^- and the τ data and hence a definite conclusion is not possible. Nevertheless, the above considerations point out that improved agreement between the world average value of α_s and the value of α_s determined from the low-energy e^+e^- data, may be achieved if the e^+e^- spectral densities become larger; this would automatically entail the increase in the theoretical value of the muon anomalous magnetic moment and a smaller disagreement with the experimental value.

The two examples discussed above lead to *contradictory* conclusions in as much as the consistency of the muon anomalous magnetic moment with other precision data is concerned. While the Higgs mass constraint indicates that a larger value of a_μ^{th} pushes m_H lower to an extent that a conflict with the direct bound becomes a possibility, the world average value of the strong coupling constant seems to require a *larger* value of a_μ^{th}. Whether or not these contradictions point to physics beyond the Standard Model as a solution is open to debate. However, it is undeniable and intriguing that, for the last few years, the theoretical estimate of the muon magnetic anomaly is persistently

---

[2]   Values of $\alpha_s(M_Z)$ obtained from τ decays and from heavy quarkonia spectroscopy are not included when the average is computed.

lower than the experimental value. It is therefore interesting to investigate the consequences of this discrepancy, assuming that its cause is physics beyond the Standard Model. We review some of new physics explanations of the $g-2$ discrepancy in the next chapter where we show that its magnitude is natural in many new physics models.

References

1. G. Charpak, F. J. M. Farley, R. L. Garwin, T. Muller, J. C. Sens, V. L. Telegdi and A. Zichichi, Phys. Rev. Lett. **6**, 128 (1961); Nuovo Cimento **22**, 1043 (1961); Phys. Lett. **1**, 16 (1962).
2. G. Charpak, F. J. M. Farley, R. L. Garwin, T. Muller, J. C. Sens, V. L. Telegdi and A. Zichichi, Nuovo Cimento **37**, 1241 (1962).
3. F. J. M. Farley, J. Bailey, R. C. A. Brown, M. Giesch, H. Jöstlein, S. van der Meer, E. Picasso and M. Tannenbaum, Nuovo Cimento **45**, 281 (1966).
4. J. Bailey, G. von Bochmann, R. C. A. Brown, F. J. M. Farley, H. Jöstlein, E. Picasso and R. W. Williams, Phys. Lett. B **28**, 287 (1968).
5. J. Bailey, W. Bartl, G. von Bochmann, R. C. A. Brown, F. J. M. Farley, M. Giesch, H. Jöstlein, S. van der Meer, E. Picasso and R. W. Williams, Nuovo Cimento A **9**, 369 (1972).
6. J. Bailey et al., Nucl. Phys. B **150**, 1 (1979).
7. H. N. Brown et al., [Muon g-2 Collaboration], Phys. Rev. D **62**, 091101 (2000).
8. H. N. Brown et al., [Muon g-2 Collaboration], Phys. Rev. Lett. **86**, 2227 (2001).
9. G. W. Bennett et al. [Muon g-2 Collaboration], Phys. Rev. Lett. **89**, 101804 (2002); Erratum-ibid. **89**, 129903 (2002).
10. G. W. Bennett et al. [Muon g-2 Collaboration], Phys. Rev. Lett. **92**, 161802 (2004).
11. W. J. Marciano, Nucl. Proc. Suppl., **166**, 437 (2003); Lecture "Precision Electroweak Measurements and the Higgs mass," in [13].
12. K. Maltman, Phys. Lett. B **633**, 512 (2006).
13. Proceedings of 32nd SLAC Summer Institute on Particle Physics "Nature's Greatest Puzzles," J. L. Hewett, J. Jaros, T. Kamae and C. Prescott (eds), eConf C040802.

8 New Physics and the Muon Anomalous Magnetic Moment

In this chapter we review various scenarios of physics beyond the Standard Model that may explain the difference between the measured value of the muon anomalous magnetic moment and the Standard Model expectation for this quantity. It was realized long ago that the muon anomalous magnetic moment provides a powerful tool for studying the energy frontier. For example, in late 1950s, the interest in the muon magnetic anomaly was driven by a quest to understand why the muon is heavier than the electron while all other properties of the two leptons are the same. It was suggested that the muon magnetic anomaly is a particular suitable observable since it can be used to probe the validity of QED up to the energy scale of about few GeV, the energy frontier at that time.

The current situation with the muon anomalous magnetic moment is similar although, clearly, energy scales that are involved are quite different. Collider searches for beyond the Standard Model physics and good agreement between precision electroweak measurements and the Standard Model imply that new phenomena should appear at energy scales higher than $0.5 - 1$ TeV. It is expected that new particles with masses in this range will be accessible to the Large Hadron Collider (LHC) that will start operating in 2007. There is no shortage of suggestions of what the physics beyond the Standard Model might be [1], but all these ideas have to await experimental confirmation.

The muon anomalous magnetic moment constrains new physics indirectly, through contributions of virtual, yet undiscovered particles. If an allowed range of the difference $a_\mu^{\mathrm{BSM}} = a_\mu^{\mathrm{exp}} - a_\mu^{\mathrm{th}}$ is established, it can be used to constrain the parameter space of a particular realization of the beyond the Standard Model physics. Because the muon anomalous magnetic moment is a single low-energy observable, it can not be used to uniquely identify the physics beyond the Standard Model, but this can probably be achieved in a combination with other low-energy observables and direct searches at colliders.

Shortly after first results of the Brookhaven experiment E821, announced in 2001, indicated a possible disagreement between a_μ^{exp} and a_μ^{th}, various models of beyond the Standard Model physics were invoked to explain it. These models include supersymmetry [2, 3], extensions of the Higgs sector of the Standard Model [4], models with additional gauge bosons, such as left-

K. Melnikov and A. Vainshtein: *Theory of the Muon Anomalous Magnetic Moment*
STMP **216**, 151–170 (2006)
© Springer-Verlag Berlin Heidelberg 2006

right symmetric models [5], compositeness, extra dimensions [6, 7, 8, 9], and others. In what follows, we briefly review some of these explanations.

Before doing so, it is useful to fix the allowed new physics contribution to a_μ, consistent with the current data and the theoretical calculations of the muon magnetic anomaly in the Standard Model. In this chapter, we assume that at the 95% confidence level, the new physics contribution should fall into the interval (cf. (7.5))

$$40 \times 10^{-11} < a_\mu^{\mathrm{BSM}} < 440 \times 10^{-11} \ . \tag{8.1}$$

We note that for constraining new physics, it is crucial that a_μ^{BSM} has the *lower* bound which implies that new physics can not decouple. Unfortunately, such a conclusion is only valid if, as in (8.1), we consider the 95% confidence level (2σ) interval; when the 3σ interval is considered, the lower bound is negative and $a_\mu^{\mathrm{BSM}} = 0$ becomes consistent with the data.

8.1 Supersymmetry

Supersymmetry (SUSY) extends in a non-trivial fashion space-time symmetries (the Poincare group of translations and rotations in Minkowski space) to include generators that relate bosons and fermions. Its practical consequence is rather curious – every particle P must have a superpartner $\tilde{P}$ whose spin differs from that of P by one half and whose mass equals to that of P. In particular, in a supersymmetric world, there would be such scalar particles as *sleptons* and *squarks*, partners of ordinary quarks and leptons, partners of gauge bosons *gauginos* and others. Such particles are not observed in Nature; hence, if the supersymmetry has anything to do with the real world, it has to be broken in such a way that superpartners of all known elementary particles are sufficiently heavy to be unobservable at existing colliders. Nevertheless, if indeed present in Nature, superpartners of ordinary particles may contribute to low-energy observables through quantum loops. Because the muon anomalous magnetic moment is one of the most precisely measured low-energy observables, it might be a good place to detect superpartners. We discuss implications of supersymmetry for the muon anomalous magnetic moment in Sect. 8.1.2. Before that, we describe peculiar implications of the unbroken supersymmetry for the anomalous magnetic moment of an elementary fermion in supersymmetric extension of the U(1) gauge theory.

8.1.1 The Anomalous Magnetic Moment
in Supersymmetric QED

Consider supersymmetric version of QED (SQED). For the sake of argument, we assume that the theory only contains muons as matter fermions. The muon (μ^-) has two spin states; then, supersymmetry requires two bosons, smuons, as superpartners. One smuon, S, is a scalar, another, P, is a pseudoscalar. A superpartner of photon, photino, is represented by the Majorana fermion field λ_α. The Lagrangian of the theory reads [10]

$$\mathcal{L}_{\text{SQED}} = \mathcal{L}_{\text{QED}} + \mathcal{L}_{\text{SUSY}} , \qquad (8.2)$$

where

$$\mathcal{L}_{\text{QED}} = -\frac{1}{4} F_{\mu\nu} F^{\mu\nu} + \bar{\psi}(i\hat{D} - m)\psi , \qquad (8.3)$$

with $D_\mu = \partial_\mu - ieA_\mu$, is the ordinary QED Lagrangian and

$$\mathcal{L}_{\text{SUSY}} = \frac{1}{2} \bar{\lambda} i\hat{\partial} \lambda + |D_\mu S|^2 + m^2 |S|^2 + |D_\mu P|^2 + m^2 |P|^2$$

$$- ie \left[\bar{\psi}(S + i\gamma_5 P)\lambda - \bar{\lambda} \left(S^+ - i\gamma_5 P^+ \right) \psi \right] - \frac{e^2}{2} \left(S^+ P - SP^+ \right)^2 .$$

$$(8.4)$$

describes additional terms required to make $\mathcal{L}_{\text{SQED}}$ supersymmetric. Note that as the consequence of supersymmetry, all couplings in the theory are defined through the muon electric charge e.

The supersymmetric generalization of QED has many interesting properties. For example, supersymmetry imposes strict restrictions on the renormalization properties of the theory; in general, ultra-violet properties of supersymmetric theories are better, than in ordinary gauge theories. The generic reason for this is the cancelation between Feynman graphs present in ordinary QED with "new" Feynman graphs that involve superpartners.

The consequences of these fact are numerous. For example, it turns out possible to compute an *exact* β-function for such theories [11]. Also, masses of scalar particles become stable under radiative corrections; this means, that the quadratic divergence that is present in the self-energy operator of a scalar particle in ordinary QED disappears in the supersymmetric version of the theory. It is interesting to point out that this feature of the theory is exploited in an attempt to extend the Standard Model in a way that allows to solve the so-called hierarchy problem.

Another feature of SQED is, perhaps, less known but it is quite interesting as well. It was pointed out by Ferrara and Remiddi that in SQED the matter fermion (muon, in our case) can not acquire the anomalous magnetic moment [12]. They proved this fact by demonstrating that the magnetic dipole moment operator, $F_{\mu\nu}\bar{\psi}\sigma^{\mu\nu}\psi$ is not consistent with SUSY. The result is valid through all orders in the coupling constant. We will not reproduce here the general proof, limiting ourselves to a demonstration of the cancelation at the one-loop level.

From the Lagrangian of SQED (8.2) it is clear that there are three diagrams that contribute to the muon anomalous magnetic moment. The first one is the standard QED contribution studied in Chap. 2. The other two originate from the muon-smuon-photino vertices in $\mathcal{L}_{\text{SUSY}}$. Such diagrams are shown in Fig. 8.1a, where $\tilde{\mu} = P$ or S and $\chi^0 = \lambda$. The vertex functions (cf. (2.3)) for the two diagrams are

$$\Gamma^{S,\mu} = ie^2 \int \frac{\mathrm{d}^4 k}{(2\pi)^4} \frac{\hat{k} \, (p_2^\mu + p_1^\mu - 2k^\mu)}{k^2((p_2 - k)^2 - \tilde{m}^2)((p_1 - k)^2 - \tilde{m}^2)} ,$$

$$\Gamma^{P,\mu} = -ie^2 \int \frac{\mathrm{d}^4 k}{(2\pi)^4} \frac{\gamma_5 \hat{k} \gamma_5 \, (p_2^\mu + p_1^\mu - 2k^\mu)}{k^2((p_2 - k)^2 - \tilde{m}^2)((p_1 - k)^2 - \tilde{m}^2)} ,$$

$$(8.5)$$

where $\tilde{m}$ stands for the smuon mass that we treat as a free parameter for the time being. Of course, exact supersymmetry requires $\tilde{m} = m$.

Using $\gamma_5 \hat{k} \gamma_5 = -\hat{k}$, we observe that the two vertex functions are identical. To compute the smuon contribution to the muon anomalous magnetic moment, we follow the calculation discussed in Sect. 2.3. Neglecting terms that do not contribute to the Pauli form factor, we derive

$$\Gamma^{\tilde{\mu},\mu} = \Gamma^{S,\mu} + \Gamma^{P,\mu} \stackrel{a_\mu}{=} \frac{\alpha}{4\pi} \frac{(p_2^\mu + p_1^\mu)}{2m} G(\tilde{m}/m) , \qquad (8.6)$$

where

$$G(\tilde{m}/m) = 2 \int\limits_0^1 \mathrm{d}x \, \frac{x(1 - x)}{(\tilde{m}/m)^2 - x} . \qquad (8.7)$$

Using the identity

$$\bar{u}_{p_2} \gamma^\mu u_{p_1} = \bar{u}_{p_2} \left[\frac{p_2^\mu + p_1^\mu}{2m} + \frac{i\sigma^{\mu\nu} q_\nu}{2m} \right] u_{p_1} , \qquad (8.8)$$

where $q = p_2 - p_1$, we read off the smuon contribution to the muon anomalous magnetic moment from (8.6). We find

$$a_\mu^{\tilde{\mu}} = -\frac{\alpha}{2\pi} G(\tilde{m}/m) . \qquad (8.9)$$

The muon anomalous magnetic moment in SQED in the one-loop approximation is given by the sum of $a_\mu^{\tilde{\mu}}$ and the Schwinger correction (2.27),

$$a_\mu^{\mathrm{SQED}} = \frac{\alpha}{2\pi} \left(1 - G(\tilde{m}/m)\right) . \qquad (8.10)$$

The exact supersymmetry requires the equality of the muon and smuon masses. Since $G(1) = 1$, we find that $a_\mu^{\mathrm{SQED}} = 0$ in case when the supersymmetry is unbroken.

An absence of the anomalous magnetic moment is a particular example of numerous SUSY constraints. Generalizations [12, 13] include vanishing of the Pauli form factor $F_P(q^2)$ not only at $q^2 = 0$ but at arbitrary q^2, as well as specific relations for higher electric and magnetic multipole moments of elementary particles with spin $s > 1/2$. Further discussion of this issue requires a working knowledge of supersymmetric gauge theories and, for this reason, we do not pursue it here.

What happens when the smuon and muon masses are different? As we mentioned already, supersymmetry *has* to be broken if supersymmetric theories are to remain phenomenologically viable; moreover $\tilde{m} \gg m$ is required by existing experimental bounds. In that limit, $G(z) \approx 1/(3z^2)$, so that the contribution of smuons to the muon anomalous magnetic moment reads

$$a_\mu^{\tilde{\mu}} \approx -\frac{\alpha}{6\pi}\frac{m^2}{\tilde{m}^2} \, . \tag{8.11}$$

Although the sign of $a_\mu^{\tilde{\mu}}$ is opposite to what is required for a_μ^{BSM}, (8.1), it is interesting to estimate the magnitude of $a_\mu^{\tilde{\mu}}$. To have $|a_\mu^{\tilde{\mu}}| \approx 200 \times 10^{-11}$ we need $\tilde{m} \approx 30$ GeV. Smuons with the mass 30 GeV are excluded phenomenologically; the current lower bound on slepton masses is $\tilde{m} \gtrsim 100$ GeV. Nevertheless, it is interesting to observe that even in simplest supersymmetric extension of QED, the masses of superpartners required to obtain a_μ^{BSM} (8.1) are reasonably close to current exclusion limits. As we discuss in Sect. 8.1.2, more sophisticated supersymmetric models are totally capable of explaining the current discrepancy in the muon magnetic anomaly and, simultaneously, escape all other phenomenological constraints.

8.1.2 Supersymmetric Extensions of the Standard Model

In supersymmetric extensions of the Standard Model, the muon anomalous magnetic moment receives contributions from two diagrams, one with smuon and neutralino and the other with chargino and sneutrino, Fig. 8.1. In principle, there is also a new contribution due to modification of the Higgs sector in supersymmetric extensions of the Standard Model, but it is negligible given existing constrains on the Higgs boson masses. The Higgs contribution to a_μ may become important once the mass constraints are lifted; this is possible in the two-Higgs doublet model that is described in Sect. 8.4.

For a general supersymmetric model, a_μ^{susy} depends on seven parameters; gaugino masses $M_{1,2}$ for two electroweak gauge groups $U_Y(1)$ and $SU_L(2)$, the ratio of vacuum expectation values of up and down Higgs doublets $\tan\beta = v_u/v_d$, the μ-parameter ($\mu H_u H_d$ term in the superpotential), A_μ – the coefficient of the relevant Yukawa term in the soft SUSY breaking Lagrangian and $m_{\mathrm{L,R}}$, the masses of left- and right-handed smuons. Calculation of a_μ^{susy} in a general framework of the Minimal Supersymmetric Standard Model (MSSM) was performed in [14]. Such computation requires diagonalizing mass matrices for smuons and gauginos; as a consequence, the general result for a_μ^{susy} is somewhat convoluted. However, it is easy to display its important features and to understand the impact that precise measurement of the muon anomalous magnetic moment has on restricting SUSY parameter space.

In the context of supersymmetric contributions to the muon anomalous magnetic moment, it was realized early on that a_μ^{susy} scales as $\tan\beta$, when $\tan\beta$ is large [15]. Henceforth, for large $\tan\beta$, the supersymmetric contributions to the muon anomalous magnetic moment can become large and the

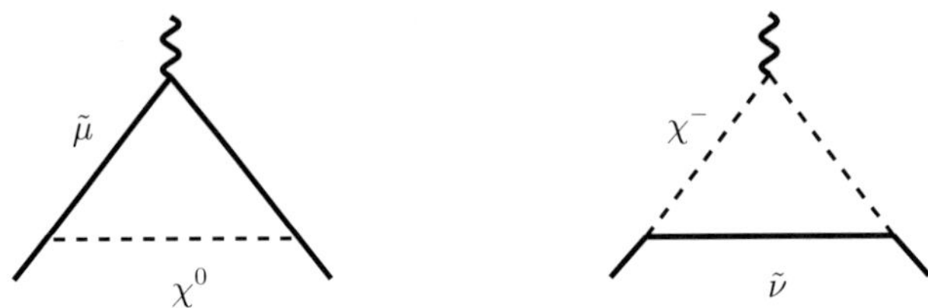

Fig. 8.1. Supersymmetric contributions to the muon anomalous magnetic moment

range, allowed for a_μ^{BSM}, can efficiently constrain such regions of the SUSY parameter space.

When $\tan\beta$ is large and masses of all sparticles are set to a common value M_{susy}, the correction to the muon anomalous magnetic moment can be written in a simple form [14]

$$a_\mu^{\mathrm{susy}} = \frac{\mathrm{sgn}(\mu)\, m^2 \tan\beta}{192\pi^2 M_{\mathrm{susy}}^2}\left(g_1^2 + 5g_2^2\right) \approx \frac{\alpha(M_Z)}{8\pi \sin^2\theta_W}\frac{\mathrm{sgn}(\mu)\, m^2 \tan\beta}{M_{\mathrm{susy}}^2}\,. \quad (8.12)$$

The dominant contribution to a_μ^{susy} is due to chargino-sneutrino diagram whereas the neutralino-smuon diagram gives a contribution that is smaller, approximately, by an order of magnitude. As expected, a_μ^{susy} scales as $1/M_{\mathrm{susy}}^2$; also, it is proportional to $\tan\beta$ and to the sign of the μ-parameter.

An immediate consequence of (8.12) is that if supersymmetry is to explain the range of a_μ^{BSM} given in (8.1), the μ parameter has to be positive. Using (8.1, 8.12), we derive a useful relation between the scale of sparticle masses M_{susy} and $\tan\beta$

$$M_{\mathrm{susy}} \approx 88\ \mathrm{GeV}\sqrt{\tan\beta\,\frac{200\times10^{-11}}{a_\mu^{\mathrm{BSM}}}}\,. \quad (8.13)$$

This equation shows the order of magnitude that sparticle masses should have if a_μ^{BSM} in the interval (8.1) is to be explained by supersymmetry. For example, (8.13) shows that for the central value of $a_\mu^{\mathrm{BSM}} = 240\times10^{-11}$, the direct LEP limit on sparticle masses $M_{\mathrm{susy}} > 100$ GeV implies an absolute lower bound $\tan\beta > 1.4$. Of course, when the full range of a_μ^{BSM} is considered, the bounds on M_{susy} and $\tan\beta$ become loose; nevertheless, as a matter of principle, they exist. The above discussion should be taken only as indicative; a more careful analysis is required to achieve a definite conclusion about the impact of the a_μ measurement on the supersymmetric parameter space.

Before presenting the results of the comprehensive analysis, we remark that an unconstrained SUSY gives a possibility to explain a_μ^{BSM} by supersymmetric contributions that are *not* enhanced by the large $\tan\beta$. An example of this situation [16] is the limit where the charginos are decoupled and the neutralinos are bino-like. This can be realized by assuming $M_1 \sim m_{\tilde{\mu}} \ll M_2, \mu$. In this approximation, the chargino-sneutrino diagram decouples and the major contribution to a_μ^{susy} comes from smuon-neutralino diagram that gives

$$a_\mu^{\text{susy}} \sim \frac{\alpha(M_Z)}{12\pi \sin \theta_W^2} \frac{\text{Re}\left[\mu \tan \beta - A_1\right] m^2 M_1}{m_{\tilde\mu}^4} . \qquad (8.14)$$

Numerically, this contribution can be quite large, to allow for $a_\mu^{\text{susy}} \sim 200 \times 10^{-11}$ even for moderate $\tan \beta$, provided that the smuon masses are not very large. However, such solution is not always possible since in specific model of supersymmetry breaking the gaugino masses are proportional to each other and the chargino contribution can not be decoupled without decreasing the neutralino-smuon contribution to an unacceptably small value.

To analyze what the allowed range for a_μ^{BSM} (8.1) implies for supersymmetric parameter space, one can either scan over the seven-dimensional subspace of the general supersymmetric parameter space or consider specific models of supersymmetry breaking, for example minimal supergravity (mSUGRA), where the supersymmetry breaking is introduced in such a way that the number of independent parameters is reduced. In addition to reducing the dimensionality of the parameter space, the relations between different supersymmetric parameters lead to stronger correlations of a_μ with other low and high energy observables. For example, since a_μ is not directly sensitive to squarks and gluinos, in unconstrained MSSM there is little correlation between the muon anomalous magnetic moment and the branching ratio of $b \to s + \gamma$. However, in mSUGRA a strong correlation appears because at the scale of supersymmetry breaking, there is a single mass parameter for all scalars (squarks and sleptons) and another single mass parameter for gauginos (gluinos, binos, winos). We will illustrate the two possible approaches to the analysis of the supersymmetric contributions to the muon magnetic anomaly by considering the unconstrained SUSY [17] and the mSUGRA [18] cases.

In [17], the analysis is performed under the following assumptions: i) all supersymmetric parameters relevant for the muon magnetic anomaly are real; ii) $|\mu| > M_2$; iii) the gaugino masses satisfy $M_1 = 0.5 \, M_2$; iv) the scalar cubic coupling of smuons to the Higgs boson is bounded from above $|A_\mu| < 3|m_{\tilde\mu}|$ and v) the smuon masses are larger than the direct LEP limit [20] $m_{\tilde\mu} > 95$ GeV. It is also assumed in [17] that $-360.8 \times 10^{-11} < a_\mu^{\text{susy}} < 860 \times 10^{-11}$ which, at the time when [17] was written, used to be a 5σ allowed region (currently, it corresponds to the 6σ region). It is interesting that even such loose bounds on a_μ^{susy} in a fairly general SUSY framework lead to exclusion of some regions of the parameter space that no other observable or experiment has probed so far.

In Fig. 8.2 the exclusion regions for various values of $\tan \beta$ are shown [17]; the SUSY breaking parameters satisfy the relations displayed above. The constraint from the muon magnetic anomaly affects sparticle masses of about few hundred GeV, in accord with the estimate given in (8.13). Note,

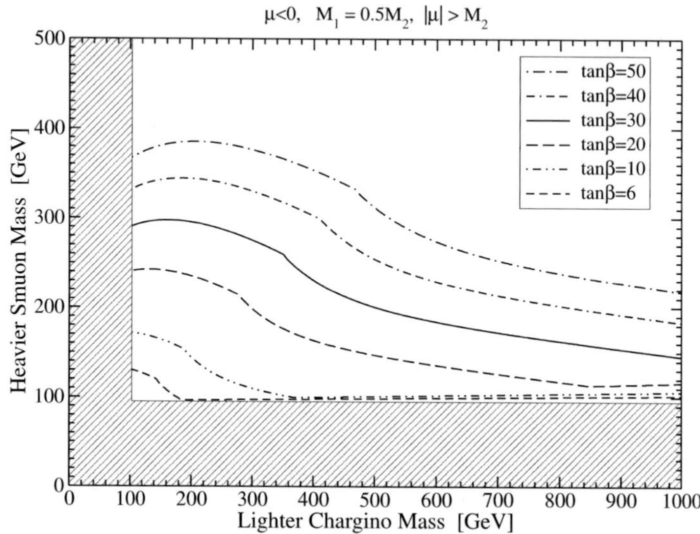

Fig. 8.2. The exclusion plot from for $\mu < 0$ and other parameters as described in the text. The shaded region is excluded by direct LEP limit. Reprinted with permission from [17] (Copyright (2003) by the American Physical Society)

that from a_μ alone, there is no upper bound on chargino masses; this part of the parameter space is excluded because for large chargino masses, one of the smuons becomes lighter than the direct bound from LEP. As emphasized in [17], Fig. 8.2 demonstrates the power of the muon magnetic anomaly in constraining supersymmetric parameter space.

The impact of other constraints on the exclusion region is shown in Fig. 8.3. Introduction of mass constraints for staus and neutralinos affects the exclusions regions. As expected, the exclusion regions become larger but imposed constraints mostly affect the region of the parameter space where charginos are heavy; for relatively light charginos, the constraint from the muon magnetic anomaly remains important.

In the framework of the unconstrained SUSY that we just discussed, the correlation between the muon anomalous magnetic moment and other low-energy processes, potentially sensitive to supersymmetry, are not significant [17] and can be evaded. The situation changes if a specific model of SUSY breaking is introduced. A particular example is the mSUGRA model where, at the Planck scale, all scalars have a common mass m_0, all gauginos have a common mass $m_{1/2}$ and all trilinear soft SUSY breaking couplings have a common value A_0. As a consequence, the model has five parameters m_0, $m_{1/2}$, A_0, $\tan\beta$, $\mathrm{sgn}(\mu)$. The mass spectrum of supersymmetric particles is computable from this set of parameters and supersymmetric contributions to processes that, at first sight, are not related, exhibit significant correlations. To demonstrate that, in Fig. 8.4 we display the plot from [18], where

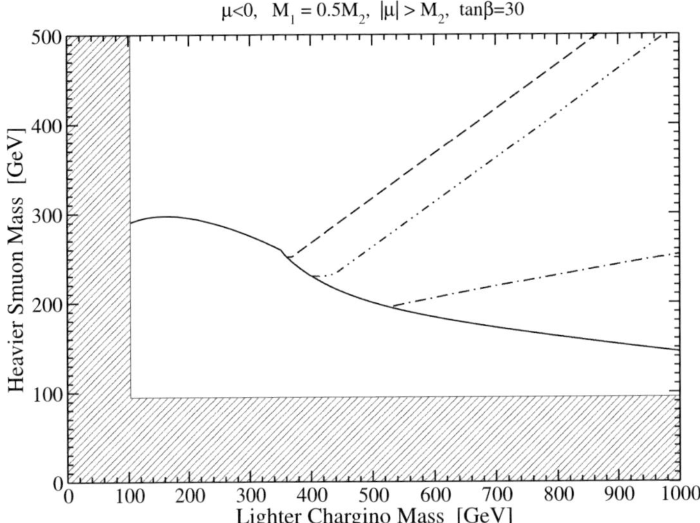

Fig. 8.3. The exclusion plot from for $\mu < 0$, $\tan\beta = 30$ and other parameters as described in the text. The excluded areas obtained by adding requirements that the stau mass is larger than 80 GeV (*dash-dotted*); the neutralino is lighter than the smuon (*dash-dot-dotted*); the neutralino is the lighter than stau (*dash*). Reprinted with permission from [17] (Copyright (2003) by the American Physical Society)

allowed regions in $(m_0, m_{1/2})$ plane for a particular choice of $A_0, \tan\beta$ and $\mathrm{sgn}(\mu)$ are shown.

In this plot contours for fixed $b \to s + \gamma$ branching fraction, a_μ^{susy}, neutralino cold dark matter density and the LEP limit on the Standard Model-like Higgs boson mass $m_h > 114\,\mathrm{GeV}$ [21] are shown. The light-shaded regions are excluded because either proper electroweak symmetry breaking is absent or the neutralino is not the lightest supersymmetric particle. The LHC reach in a generic channel $pp \to E_T + \text{missing}$ energy is also displayed. Current limits on the input values are $2.94 \times 10^{-4} < \mathrm{Br}(B \to X_s + \gamma) < 4.47 \times 10^{-4}$ [22] and $0.094 < \Omega_{\tilde{Z}_1} h^2 < 0.129$ [23]; the allowed supersymmetric contribution to a_μ is given in (8.1).

We see that the allowed region of scalar and gaugino masses for the particular choice of other supersymmetric parameters receives non-trivial constraints from all of the observables. From Fig. 8.4 we may conclude that, for this choice of $\tan\beta, \mathrm{sgn}(\mu)$ and A_0, there are two small regions in the parameter space consistent with the current data; either

$$500\,\mathrm{GeV} < m_{1/2} < 1\,\mathrm{TeV}\,, \qquad 250\,\mathrm{GeV} < m_0 < 1\,\mathrm{TeV}\,,$$

or

$$200\,\mathrm{GeV} < m_{1/2} < 400\,\mathrm{GeV}\,, \qquad 1\,\mathrm{TeV} < m_0 < 2.5\,\mathrm{TeV}\,.$$

As follows from Fig. 8.4, both of these regions in the parameter space can be explored at the LHC.

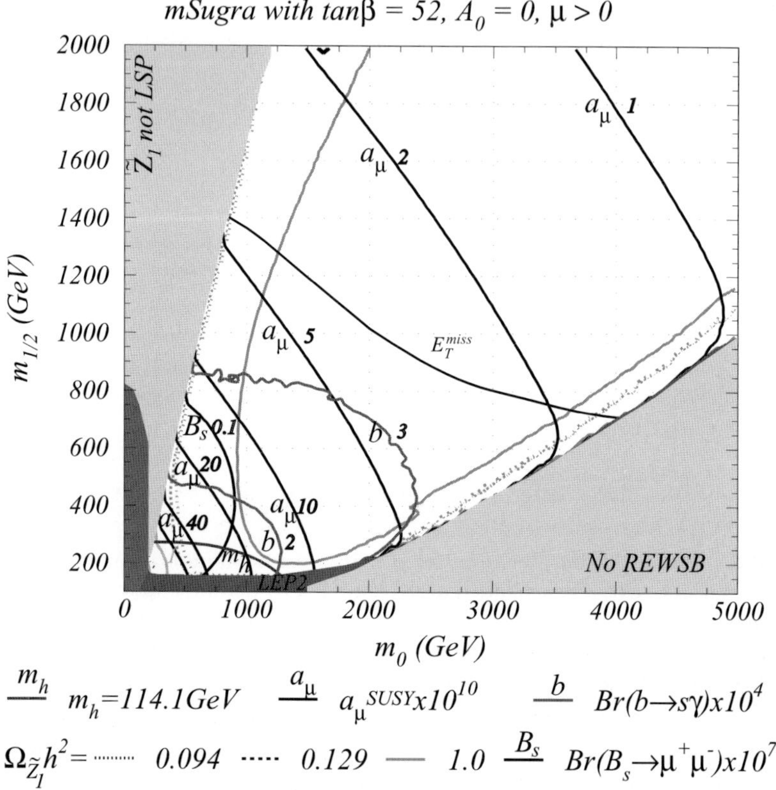

Fig. 8.4. Various constraints on mSUGRA for $\mu > 0$, $A_0 = 0$ and $\tan\beta = 52$ [18]. The shaded regions are excluded by either direct LEP limit or neutralino not being the lightest supersymmetric particle or improper electroweak symmetry breaking

To summarize, the impact of the E821 measurement on supersymmetry is significant. A simple estimate of the supersymmetric contribution to the muon magnetic anomaly (8.13) shows that sparticle masses of about few hundred GeV can account for a_μ^{BSM} in (8.1). Even when very loose bounds on allowed a_μ^{susy} are assumed, the muon anomalous magnetic moment cuts away a significant region of the parameter space in a *general* supersymmetric model; this, in particular, concerns the region of relatively light chargino and moderately light smuons. When popular supersymmetric models such as mSUGRA etc. are considered, the situation is more complex since in such models relations between different, a priori independent parameters exist at the scale where the supersymmetry is broken. As a consequence, the supersymmetric parameter space becomes constrained by a number of low-energy and collider observables and the interplay between them can be quite intricate. Simplest models of supersymmetry breaking, e.g. mSUGRA, are tightly constrained; regions of the supersymmetric parameter space that are still allowed by the existing data, are within the reach of the LHC.

8.2 Additional Vector Bosons

Recall, that exchanges of Z and W provide the one-loop electroweak correction to the muon anomalous magnetic moment; if additional gauge bosons are present in the theory, they also contribute to a_μ [5]. The appearance of gauge bosons heavier than Z and W is generic for many extensions of the Standard Model. A particular example is provided by the so-called left-right symmetric models [24], that extend the $SU(2)_L \times U(1)$ Standard Model gauge group to the $SU(2)_R \times SU(2)_L \times U(1)$. In the left-right symmetric model, there are four charged and two neutral gauge bosons of the non-abelian electroweak gauge groups and the $U(1)$ gauge boson. The electroweak symmetry is broken by the Higgs mechanism in such a way that right- and left-handed gauge bosons acquire different masses. In general, in such a model the muon couplings to gauge bosons and the Higgs boson differ from Standard Model couplings; this generates new contributions to the muon magnetic anomaly.

Given existing constraints on the masses of additional gauge bosons from collider experiments, their contribution to the muon magnetic anomaly is typically smaller than $a_\mu^{\mathrm{BSM}} \sim 200 \times 10^{-11}$. Indeed, an additional (right-handed) gauge boson with the mass M_{W_R} and the coupling to muon g_R, changes the muon anomalous magnetic moment by

$$a_\mu^{\mathrm{new}} = \frac{10}{3} \frac{G_F m^2}{8\sqrt{2}\pi^2} \frac{g_R^2 m_W^2}{g_L^2 m_{W_R}^2} . \tag{8.15}$$

Taking $a_\mu^{\mathrm{new}} > 40 \times 10^{-11}$, to conform to (8.1), we find $M_{W_R} < 400\, g_R/g_L$ GeV. However, from direct searches at the Tevatron Run I, $M_{W_R} > 720$ GeV. To comply with this bound, we require $g_R > 2g_L$, which is not very appealing theoretically since it destroys the left-right symmetry from the start.

A similar situation occurs for the contribution of an additional neutral gauge bosons to the muon magnetic anomaly. Such gauge bosons are generically referred to as Z'. $a_\mu^{\mathrm{BSM}} \approx 200 \times 10^{-11}$ requires $M_{Z'} \sim 100$ GeV, provided that the Z' coupling to muons is Standard Model-like. However, direct limits from the Tevatron Run I constrain the Z' mass $M_{Z'} > 500 - 600$ GeV. Thus, it appears that existing experimental bounds on additional gauge bosons imply that their contribution to a_μ is at the level of 10×10^{-11}. It is unlikely that the theoretical uncertainty on the calculation of the muon magnetic anomaly in the Standard Model will ever reach that value; hence, the muon anomalous magnetic moment does not seem to offer a competitive constraint on models with additional gauge bosons.

8.3 Compositeness of the Standard Model Gauge Bosons

In the Standard Model, the self-interaction of gauge bosons is fixed by the renormalizability of the theory. However, in models where gauge bosons are composite, such as the technicolor, their interaction may deviate from the canonical form. In particular, the $\gamma W^+ W^-$ effective Lagrangian can be written as

$$\mathcal{L}^{\gamma WW} = -ie\Big\{ A^\mu (W^-_{\mu\nu} W^{+\nu} - W^+_{\mu\nu} W^{-\nu})$$
$$+ \kappa F^{\mu\nu} W^+_\mu W^-_\nu + \frac{\lambda}{m^2_W} F^{\mu\nu} W^{+\rho}_\mu W^-_{\rho\nu} \Big\} , \tag{8.16}$$

where $F_{\mu\nu} = \partial_\mu A_\nu - \partial_\nu A_\mu$, $W_{\mu\nu} = \partial_\mu W_\nu - \partial_\nu W_\mu$. Two parameters, κ and λ, define the magnetic dipole moment

$$\mu_W = (1 + \kappa + \lambda) \frac{e}{2m_W}$$

and the electric quadrupole moment

$$q_W = -(\kappa - \lambda) \frac{e}{m^2_W}$$

of the W boson. In the Standard Model, $\kappa = 1$ and $\lambda = 0$.

When the gauge boson interactions are modified, the theory becomes non-renormalizable; as a consequence, the one-loop contribution to the muon magnetic anomaly becomes divergent. Assuming that loop integrals are cut off at the scale of compositeness Λ, we obtain [25]

$$a^{\text{new}}_\mu = \frac{G_F m^2}{4\sqrt{2}\pi^2} \left((\kappa - 1) \ln \frac{\Lambda^2}{M^2_W} - \frac{\lambda}{3} \right). \tag{8.17}$$

The couplings κ and λ are constrained by the LEP data on the process $e^+ e^- \to W^+ W^-$. At the 68% confidence level these constraints are [26]

$$\kappa - 1 = -0.016^{+0.042}_{-0.047} , \qquad \lambda = -0.016^{+0.021}_{-0.023} . \tag{8.18}$$

To maximize a^{new}_μ in (8.17), we take the maximal value for $\kappa - 1$ and the minimal value for λ. Then, for $\Lambda = 1$ TeV, we obtain

$$\left[a^{\text{new}}_\mu \right]_{\text{max}} = 34 \times 10^{-11} . \tag{8.19}$$

Note that the effect of λ is very small, it is only 10% of the above number.

Hence, we conclude that anomalous gauge boson couplings is an unlikely explanation of the current discrepancy in the muon magnetic anomaly.

8.4 Extensions of the Higgs Sector

The Higgs boson is the only Standard Model particle that has not yet been observed and whose couplings to gauge bosons and fermions have not been experimentally verified. Because of that, it is interesting to ask if the muon anomalous magnetic moment can be used to constrain the Higgs sector of the yet undiscovered fundamental theory.

In the Standard Model, the Higgs boson coupling to a fermion of mass m_f is $\sim m_f/m_W$ times smaller than standard electroweak couplings. Given that the one-loop electroweak correction to a_μ is, approximately, 200×10^{-11}, and the mass of the Standard Model Higgs boson is limited by LEP searches, $m_H > 114$ GeV, the contribution of the Standard Model Higgs boson to a_μ is at the level of 10^{-14} and is completely negligible.[1] The Higgs sector can significantly contribute to a_μ if its coupling to muons is enhanced and its mass is lowered, compared to the Standard Model.

This can be achieved in the model with two Higgs doublets, the two-Higgs doublet model (2HDM) [27]. We will discuss the so-called type two 2HDM that has two Higgs doublets that couple to up- and down-type fermions, respectively. In this regard, the Higgs sector of the 2HDM(II) is identical to that of the MSSM. Exactly as in the MSSM, after electroweak symmetry breaking, the Higgs sector of the theory consists of two scalars, one pseudoscalar and two charged Higgs bosons. In contrast to MSSM, however, the Higgs masses are free parameters and can be changed at will unless constrained by the data.

In principle, all the Higgs bosons of the 2HDM contribute to a_μ; however, the only possibility to get sufficiently large contribution is related to the scalar and pseudoscalar Higgs bosons which are allowed to be light and whose couplings to charged leptons and down-type quarks are enhanced by $\tan\beta$. We thus concentrate on the 2HDM(II), where either h or A is light, $\tan\beta$ is significant and the mixing angle in the CP-even part of the Higgs sector α is equal to β.

The one-loop correction to the muon anomalous magnetic moment from the Higgs sector in the 2HDM(II) is shown in Fig. 8.5. It reads

$$a_\mu^{\text{1loop}} = \frac{G_F m^2}{4\pi^2\sqrt{2}} \tan^2\beta \sum_{i=h,A} z_i F_i(z_i) \,, \tag{8.20}$$

where $z_{h,A} = m^2/m_{h,A}^2$ and

$$F_i(z) = \int_0^1 \mathrm{d}x \, \frac{N_i(x)}{1-x+xz^2} \,, \tag{8.21}$$

---

[1] This conclusion changes at the two-loop level where a contribution proportional to a single Higgs-muon Yukawa coupling appears, Fig. 8.6.

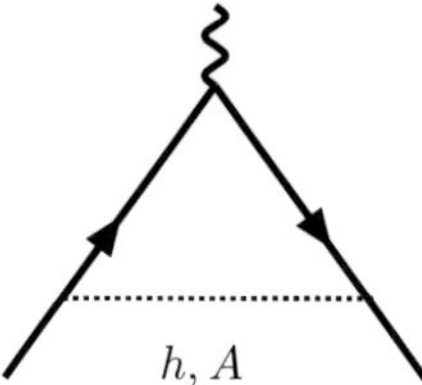

Fig. 8.5. One-loop corrections to the muon anomalous magnetic moment in 2HDM

with $N_h = x^2(2-x)$, $N_A = -x^3$. It follows from those formulas that the one-loop contribution to the muon magnetic anomaly is positive for the scalar and negative for the pseudoscalar Higgs boson. Since a_μ^{BSM}, (8.1), is positive, this seems to imply that no light pseudoscalar in the 2HDM model is allowed, while tight constraints on the scalar can be derived. A detailed analysis along these lines has been presented in [28].

However, as was pointed out in [29], one may expect significant modifications of the one-loop result when higher order corrections are included. The reason is the following. The Higgs contribution to the magnetic anomaly is small because it is suppressed by the square of the Higgs-muon Yukawa coupling. However, at the two-loop level it is possible to overcome this suppression, thanks to the diagram shown in Fig. 8.6. In this case, if the fermion in the loop is heavy, we trade the Higgs-muon Yukawa coupling for the product of the loop suppression factor α/π and the enhancement factor m_f^2/m^2, where m_f is the mass of the heavy fermion in the loop. Note, that the Yukawa coupling to up-type quarks is suppressed by $1/\tan\beta$; as a consequence, the top quark contribution is not dominant in spite of the fact that the top quark mass is very large. The largest contribution comes from the bottom quark and the tau lepton loops. Since, for example, $(\alpha/\pi)m_b^2/m^2 \sim 4$, the enhancement due to a larger Yukawa coupling overcomes the loop suppression factor and the two-loop contributions become larger than the one-loop ones.

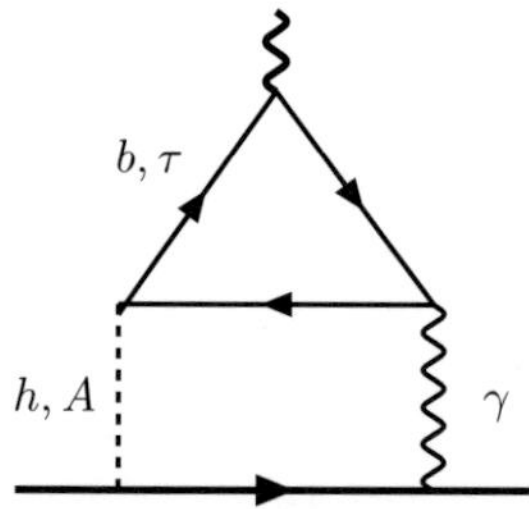

Fig. 8.6. The two-loop diagram of the Barr–Zee type [33] that gives an enhanced contribution to a_μ

The two-loop correction to the muon anomalous magnetic moment reads

$$a_\mu^{2\text{loop}} = \frac{2\alpha}{\pi} \frac{G_F m^2}{4\pi^2\sqrt{2}} \tan^2\beta \sum_{f;\, i=h,A} N_f Q_f^2 \tilde{z}_{if} \tilde{F}_{if}(\tilde{z}_{if}) , \qquad (8.22)$$

where $\tilde{z}_{hf,Af} = m_f^2/m_{h,A}^2$, $N_f = 3(1)$, for quarks and leptons, respectively, Q_f is the electric charge of the fermion and

$$\tilde{F}_{if}(z) = \int\limits_0^1 \mathrm{d}x \, \frac{\tilde{N}_i}{x(1-x)-z} \ln\frac{x(1-x)}{z} , \qquad (8.23)$$

with $\tilde{N}_h = -1 + 2x(1-x)$ and $\tilde{N}_A = 1$. We note that, in the two-loop case, the scalar contribution is negative and the pseudoscalar contribution is positive, contrary to the one-loop result discussed above. Numerical calculations show that the two-loop result is larger than the one-loop result for $m_h \geq 3$ GeV and $m_A \geq 5$ GeV. Given the signs of the one- and two-loop contributions for h and A discussed above, this fact allows to understand the influence of the muon magnetic anomaly on the 2HDM.

A comprehensive analysis of the impact of the muon anomalous magnetic moment on the parameter space of the type II 2HDM can be found in [30, 31, 32]. The main conclusion of these analyses is that a very large part of the parameter space for the 2HDM with light $m_{A,h}$ is excluded by the data. In particular, the LEP collaborations constrain the two-Higgs doublet model with light h and A since $e^+e^- \to Zh$, $e^+e^- \to Ah$ and $e^+e^- \to \bar{b}bh(A)$ processes have not been observed [34]. In particular, the OPAL data excludes large part of the parameter space where *both* h and A are light; roughly, the exclusion region can be approximated by $m_A + m_h > 90$ GeV. For our purposes, this result implies that we may consider scenarios were either the scalar or the pseudoscalar is light.

Consider the situation when the scalar is light. Then, since $a_\mu^{\text{BSM}} > 0$, this is only possible if the scalar contribution is dominated by the one-loop result and, hence, $m_h \leq 3$ GeV with $\tan\beta \geq 1$. However, such Higgs bosons are excluded by the non-observation of the Wilczek process $\Upsilon \to H + \gamma$ [35]. Hence, we conclude that the 2HDM with the light scalar Higgs boson is excluded. For the case when the pseudoscalar is light, the Higgs contribution to a_μ should be dominated by the two-loop result. As a consequence, $m_A \geq 5$ GeV is required. Part of the allowed region for $m_A \leq 10$ GeV is again excluded by the Wilczek process, whereas for larger pseudoscalar masses, the LEP experiments exclude much of the allowed parameter space. What remains [31] is a small region where 25 GeV $< m_A < 70$ GeV and $25 < \tan\beta < 100$. This region can be further reduced [32] if precision electroweak constraints are applied to the 2HDM; this happens because large mass splittings of h and A implied by OPAL data and consistent with the range of a_μ^{BSM}, are in conflict with, e.g., apparent agreement of $\Gamma(Z \to b\bar{b})/\Gamma(Z)$ with its Standard Model value.

8.5 Extra Dimensions

Another interesting idea about the beyond the Standard Model physics is the suggestion that there are more than three spatial dimensions [36, 37]. The need for additional dimensions is motivated by the huge difference in the observed scale of electroweak symmetry breaking and the Planck scale. To solve this "hierarchy problem", [36] postulates that gravity is weak due to existence of large, compactified additional dimensions, where gravitons can propagate. In the simplest version of the model [36], the Standard Model fields live on a four-dimensional manifold, called the brane, in n-dimensional space. Because of the additional volume available for gravitons; the gravitational interactions are weak, unless the energy of the processes on the brane becomes comparable to the Planck scale M_* of quantum gravity in $n = 4 + \Delta$ dimensions. The value of M_* is assumed to be of the order of 1 TeV. The relation between the Newton's constant G_N that measures the strength of the gravitational interactions on the brane and the fundamental $4 + \Delta$ Planck scale M_* reads

$$4\pi G_N R^\Delta = M_*^{-2-\Delta} \,, \tag{8.24}$$

where it is assumed that additional Δ dimensions are compactified on a circle of radius R.

In all variants of extra-dimensional theories, gravity always propagates in extra dimensions, whereas other Standard Model fields may or may not do so. The simplest possibility is to assume that geometry of additional dimensions is flat [36], but generalizations of the theory allow for a non-trivial geometry [37]. A common feature of theories with compactified additional dimensions is the appearance of the so-called Kaluza-Klein states that can be thought of as excitations of the Standard Model fields that are allowed to propagate in additional dimensions. For example, the simplest version of extra-dimensional theories features massive gravitons, in addition to the massless graviton, responsible for Newtonian gravity on the brane. Masses of Kaluza-Klein states depend on the geometry of additional dimensions; when additional dimensions are compactified on a circle of radius R, the mass of the Kaluza-Klein graviton excitation is

$$m_{\boldsymbol{n}}^2 = \frac{n_1^2 + n_2^2 + \ldots + n_\Delta^2}{R^2} \,, \tag{8.25}$$

where all quantum numbers n_i are integers.

The impact of the muon magnetic anomaly on theories with extra dimensions was studied in [6, 7, 8, 9]. We discuss the simplest scenario, where the Standard Model fields are confined to the brane and gravity is allowed to propagate in Δ compactified dimensions [6]. In this case, new physics contribution to a_μ comes from Kaluza-Klein gravitons whose masses are given by (8.25). Each Kaluza-Klein mode of the graviton couples to ordinary matter through the stress-momentum tensor $T_{\mu\nu}$, with the coupling proportional to the square root of the ordinary Newton's constants. There are two interesting features of the gravitational corrections to the muon anomalous magnetic moment [6, 38]; first, for each Kaluza-Klein mode, the gravitational correction to a_μ is *finite*; second, heavier gravitons *do not decouple*. The last feature follows from the fact that the strength of the interaction between gravitons and matter grows with energy. As the result, the correction to the muon anomalous magnetic moment reads

$$a_\mu^{\text{ed}} = \frac{G_N m^2}{2\pi} \sum_{\boldsymbol{n}} w(m, m_{\boldsymbol{n}}) \,, \tag{8.26}$$

where $w(m, m_{\boldsymbol{n}})$ is the contribution to a_μ^{ed} due to $\boldsymbol{n}$-th Kaluza-Klein mode with the mass $m_{\boldsymbol{n}}$. Using (8.25), we may rewrite the sum over quantum numbers $\boldsymbol{n}$ through the integral over $m_{\boldsymbol{n}}^2$. We find

$$a_\mu^{\text{ed}} = \frac{G_N R^\Delta m^2}{2\pi} \frac{\pi^{\Delta/2}}{\Gamma(\Delta/2)} \int \mathrm{d}s\, s^{\Delta/2-1} w(m, \sqrt{s}) \,. \tag{8.27}$$

Since, as we pointed out above, $w(m, \sqrt{s}) \to \text{const} = c$, for $\sqrt{s} \to \infty$, the integral is saturated by heavy Kaluza-Klein gravitons. Cutting the integral (8.27) at $\sqrt{s} = \lambda M_*$ and using (8.24) to remove the Newton's constant, we find

$$a_\mu^{\text{ed}} = \frac{c\lambda^\Delta}{4\pi^2} \frac{\pi^{\Delta/2}}{\Delta\Gamma(\Delta/2)} \frac{m^2}{M_*^2} \,. \tag{8.28}$$

As shown in [6], $c = 5$. In addition, there is a contribution to a_μ from the so-called radion field; we neglect it here because the radion contribution is small [6].

From (8.28) we see that a_μ^{ed} depends on the unknown parameter λ that is introduced to parametrize our inability to carry out the integral over masses of Kaluza-Klein states in (8.27). This is a typical situation with calculations in theories with extra dimensions since these theories are not renormalizable and have power divergences. For estimates, we take $\lambda = 1$ and derive

$$M_* = 1 \text{ TeV} \times \begin{cases} \left(200 \times 10^{-11}/a_\mu\right)^{1/2}, & \Delta = 2; \\ \left(327 \times 10^{-11}/a_\mu\right)^{1/2}, & \Delta = 6. \end{cases} \tag{8.29}$$

How does the estimate for the fundamental Planck scale M_* shown in (8.29) compare with other constraints on M_*? For $\Delta = 2$, severe constraints come from astrophysics and cosmology [39]; they require $M_* > $ few TeV making the muon magnetic anomaly constraint irrelevant. On the other hand, for $\Delta = 6$, astrophysical constrains are quite weak and direct limits from Tevatron and LEP require $M_* \geq 600 - 700$ GeV. In such a case, the estimate from the muon magnetic anomaly (8.29) is quite competitive since, $a_\mu^{\text{BSM}} < 440 \times 10^{-11}$, implies $M_* \geq 900$ GeV.

There are two other variants of models with extra dimensions that have been suggested. The first model is the so-called universal extra dimensions where the compactification radius R is sufficiently small so that all the Standard Model particles can propagate in the bulk. The second model is the Randall-Sundrum model [37] where the geometry of extra dimensions is not flat. The muon anomalous magnetic moment in these models was studied in [9, 8, 7]. Typically, in such models the muon magnetic anomaly receives *negative* contributions which is difficult to reconcile with the currently preferred range for a_μ^{BSM}.

References

1. Good exhibition of current ideas can be found in the Proceedings of 32nd SLAC Summer Institute on Particle Physics "Nature's Greatest Puzzles," J. L. Hewett, J. Jaros, T. Kamae and C. Prescott (eds), eConf C040802.
2. J. A. Grifols and A. Mendez, Phys. Rev. D **26**, 1809 (1982);
J. Ellis, J. Hagelin and D. V. Nanopoulos, Phys. Lett. B **116**, 283 (1982);
R. Barbieri and L. Maiani, Phys. Lett. B **117**, 203 (1982);
D. A. Kosower, L. M. Krauss and N. Sakai, Phys. Lett. B **133**, 305 (1983);
T. C. Yuan, R. Arnowitt, A. H. Chamseddine and P. Nath, Z. Phys. C **26**, 407 (1984);
T. Moroi, Phys. Rev. D **53**, 6565 (1996);
M. Carena, G. F. Giudice and C. E. Wagner, Phys. Lett. B **390**, 234 (1997);
T. Ibrahim and P. Nath, Phys. Rev. D **61**, 095008 (2000);
G. Cho, K. Hagiwara and M. Hayakawa, Phys. Lett. B **478**, 231 (2000).
3. L. Everett, G. Kane, S. Rigolin and L. Wang, Phys. Rev. Lett. **86**, 3484 (2001);
S. P. Martin and J. D. Wells, Phys. Rev. D **64**, 035003 (2001);
J. L. Feng and K. Matchev, Phys. Rev. Lett. **86**, 3480 (2001);
E. A. Baltz and P. Gondolo, Phys. Rev. Lett. **86**, 5004 (2001);
U. Chattopadhya and P. Nath, Phys. Rev. Lett. **86**, 5854 (2001);
S. Komine, T. Moroi and M. Yamaguchi, Phys. Lett. B **506**, 93 (2001); Phys. Lett. B **507**, 224 (2001);
J. R. Ellis, D. V. Nanopoulos and K. A. Olive, Phys. Lett. B **508**, 65 (2001);
R. Arnowitt, B. Dutta, B. Hu and Y. Santoso, Phys. Lett. B **505**, 177 (2001);
K. Choi et al., Phys. Rev. D **64**, 055001 (2001);
J. E. Kim, B. Kyae and H. M. Lee, Phys. Lett. B **520**, 298 (2001);
K. Cheung, C. Chou and O. C. Kong, Phys. Rev. D **64**, 111301 (2001);
H. Baer et al., Phys. Rev. D **64**, 035004 (2001);
C. Chen and C. Q. Geng, Phys. Lett. B **511**, 77 (2001);
G. Cho and K. Hagiwara, Phys. Lett. B **514**, 123 (2001).
4. E. D. Carlson, S. L. Glashow and U. Sarid, Nucl. Phys. B **309**, 597 (1988);
M. Krawczyk and J. Zochowski, Phys. Rev. D **55**, 6968 (1997);
P. H. Chankowski, M. Krawczyk and J. Zochowski, Eur. Phys. J. C **11**, 661 (1999);
A. Dedes and H. E. Haber, JHEP **0105**, 006 (2001);
F. Larious, G. Tavares-Velasco and C. P. Yuan, Phys. Rev. D **64**, 055004 (2001);
D. Chang, W. Chang, C. Chou and W. Keung, Phys. Rev. D **63**, 091301 (2001);

K. Cheung, C. Chou and O. C. Kong, Phys. Rev. D **64**, 111301 (2001); Phys. Rev. D **68**, 053003 (2003).

5. J. P. Leveille, Nucl. Phys. B **137**, 63 (1978).
6. M. L. Graesser, Phys. Rev. D **61**, 074019 (2000).
7. H. Davoudiasl, J. L. Hewett and T. Rizzo, Phys. Lett. B **493**, 135 (2000).
8. K. Agashe, N. G. Deschpande and G. H. Wu, Phys. Lett. B **511**, 85 (2001).
9. B. Dobrescu and T. Appelquist, Phys. Lett. B **516**, 85 (2001).
10. J. Wess and B. Zumino, Nucl. Phys. B **70**, 39 (1974); Phys. Lett. B **49**, 5 (1974); Nucl. Phys. B **78**, 1 (1984).
11. V.A. Novikov, M.A. Shifman, A.I. Vainshtein and V.I. Zakharov, Nucl. Phys. B **229**, 381 (1983); Phys. Lett. B **166**, 329 (1986).
12. S. Ferrara and E. Remiddi, Phys. Lett. B, **53**, 347 (1974).
13. S. Ferrara and M. Porrati, Phys. Lett. B, **288**, 85 (1992); I. Giannakis, J. T. Liu and M. Porrati, Phys. Rev. D, **58**, 025009 (1998).
14. T. Moroi, Phys. Rev. D **53**, 6565 (1996);
15. J. L. Lopez, D. V. Nanopolulos and X. Wang, Phys. Rev. D **49**, 366 (1994).
16. S. P. Martin and J. D. Wells, Phys. Rev. D **64**, 035003 (2001).
17. S. P. Martin and J. D. Wells, Phys. Rev. D **67**, 015002 (2003).
18. H. Baer, C. Balazs, A. Belyaev, T. Krupovnickas and X. Tata, JHEP **0306**, 054 (2003).
19. S. Eidelman et al. [Particle Data Group], Phys. Lett. **B592**, 1 (2004).
20. See M. Schmitt, *Supersymmetry, Part II (Experiment)*, in [19].
21. See P. Igo-Kemenes, *Searches for Higgs bosons*, in [19].
22. A recent analysis is given in T. Hurth, E. Lunghi and W. Porod, Nucl. Phys. B **704**, 56 (2005).
23. D. N. Spergel et al., Astrophys. J. Suppl. **148**, 175 (2003).
24. J. C. Pati and A. Salam, Phys. Rev. D **10**, 275 (1974);
    R. N. Mohapatra and J. C. Pati, Phys. Rev. D **11**, 566 (1975);
    G. Senjanovic and R. N. Mohapatra, Phys. Rev. D **12**, 1502 (1975).
25. P. Mery, S. E. Moubarik, M. Perrottet and F. M. Renard, Zeit. f. Phys. C **46**, 229 (1990);
    F. Herzog, Phys. Lett. B **148**, 355 (1984);
    M. Suzuki, Phys. Lett. B **153**, 289 (1985);
    A. Grau and J. A. Grifols, Phys. Lett. B **154**, 283 (1985);
    M. Beccaria, F. M. Renard, S. Spagnolo and G. Verzegnassi, Phys. Lett. B **448**, 129 (1999).
26. LEP Elecroweak Working Group, *A combination of preliminary electroweak measurements and constraints on the standard model*, hep-ex/0511027.
27. J. F. Gunion, H. E. Haber, G. L. Kane and S. Dawson, *The Higgs hunter's guide*, Addison-Wesley, Reading, MA, 1990.
28. A. Dedes and H. Haber, JHEP **0105**, 006 (2000).
29. D. Chang, W. Chang, C. Chou and W. Keung, Phys. Rev. D **63**, 091301 (2001).
30. K. Cheung, C. Chou and O. C. Kong, Phys. Rev. D **64**, 111301 (2001).
31. M. Krawczyk, Acta. Phys. Polon. B **33**, 2621 (2002).
32. K. Cheung, C. Chou and O. C. Kong, Phys. Rev. D **68**, 053003 (2003).
33. S. M. Barr and A. Zee, Phys. Rev. Lett. **65**, 21 (1990); Erratum-ibid. **65**, 2920 (1990).
34. G. Abbiendi et al. [OPAL Collaboration], Eur. Phys. J. C **18**, 425 (2001); G. Abbiendi et al. [OPAL Collaboration], Eur. Phys. J. C **23**, 397 (2002); DELPHI collaboration, DELPHI 2002-037-CONF-571; ALEPH collaboration, PA13-027.

35. P. Franzini et al. [CUSB collaboration], Phys. Rev. D **35**, 2883 (1987).
36. N. Arkani-Hamed, S. Dimopoulos and G. Dvali, Phys. Lett. B **429**, 263 (1998); I. Antoniadis, N. Arkani-Hamed, S. Dimopoulos and G. Dvali, Phys. Lett. B **436**, 257 (1998); G. Shiu and S. H. H. Tye, Phys. Rev. D **58**, 106007 (1998); N. Arkani-Hamed, S. Dimopoulos and G. Dvali, Phys. Rev. D **59**, 086004 (1999).
37. L. Randall and R. Sundrum, Phys. Rev. Lett. **83**, 3370 (1999); Phys. Rev. Lett. **83**, 4690 (1999).
38. F. A. Berends and R. Gastmans, Phys. Lett. B **55**, 311 (1975).
39. For a review, see J. L. Hewett and M. Spiropulu, Ann. Rev. Nucl. Part. Phys. **52**, 397 (2002).

9 Summary

The experiment E821, completed recently at Brookhaven National Laboratory, measured the muon magnetic anomaly to an impressive precision of about one part per million. That measurement and, in particular, preliminary indication of the disagreement between the experimental result and the Standard Model prediction, stimulated a flurry of theoretical activity. As the result, the calculation of the muon magnetic anomaly within the Standard Model was scrutinized and the understanding that existing models of new physics can easily explain the current deviation, was achieved.

To better emphasize the level at which the theoretical predictions are being tested, we point out that, currently, the theory and experiment disagree on the value of the muon magnetic anomaly in two parts per million. This tiny disagreement may be a signal that the Standard Model of particle physics, as we know it, is incomplete. When translated to standard deviations, the current disagreement becomes 2.7σ and, hence, is not conclusive. As of this writing, it is unclear whether the experimental program will continue, but new ideas for measuring the muon magnetic anomaly with even higher precision are being discussed [1].

While it seems relatively straightforward to reduce the current experimental uncertainty on a_μ to $(20 - 30) \times 10^{-11}$ by collecting more data [1], it is less clear whether a similar improvement in the theoretical uncertainty is possible. The uncertainty of the CMD-2 and SND results for the cross-section $\sigma(e^+e^- \to \pi^+\pi^-)$ is dominated by systematic uncertainties, related to QED radiative corrections. In principle, this should be improvable but dramatic changes are unlikely. The radiative return measurements at BaBar and Belle may help to improve the knowledge of the e^+e^- hadronic annihilation cross-section at higher energies. It is conceivable that the error on the hadronic vacuum polarization contribution to the muon magnetic anomaly is reducible to $(30 - 40) \times 10^{-11}$ in the long run.

The error on the hadronic light-by-light scattering contribution is difficult to improve since such an improvement would necessarily entail a breakthrough in our understanding of hadronic interactions at low energies. Hence, it appears that the reduction of the combined error below $(60 - 70) \times 10^{-11}$ is unlikely. Provided that the difference between the experimental value of a_μ

K. Melnikov and A. Vainshtein: *Theory of the Muon Anomalous Magnetic Moment*
STMP **216**, 171–172 (2006)
© Springer-Verlag Berlin Heidelberg 2006

and the theoretical prediction does not change significantly, this will move the disagreement into a much more interesting range of $3-4$ standard deviations.

However, it is most likely that the final closure of the recent story of the muon anomalous magnetic moment will come from collider experiments. In 2007, the Large Hadron Collider will begin the exploration of the new energy frontier. If the current discrepancy in the muon magnetic anomaly is genuine, the LHC will find new physics responsible for it. It is entirely possible that, say, in 2010 we will look back and realize that the first hints of new physics that will have been discovered, were there already in 2004 as the puzzling discrepancy in the muon magnetic anomaly that just did not want to go away.

References

1. D. W. Hertzog, Nucl. Phys. Proc. Suppl. **144**, 191 (2005).

A Appendix

A.1 Notation

In this Appendix, we summarize the notation that are used throughout the text. We use e to denote the muon electric charge, $e = -|e|$ and m to denote the muon mass.

The Lorentz indices are denoted by Greek letters, $\mu, \nu, ... = 0, 1, 2, 3$; the three-dimensional indices by Latin letters, $i, j, k, ... = 1, 2, 3$. The spatial vectors refer to upper Lorentz indices, e.g. $x^\mu = (t, \boldsymbol{x})$. The metric tensor is $g_{\mu\nu} = \mathrm{Diag}\{1, -1, -1, -1\}$, and the totally antisymmetric tensor $\epsilon^{\mu\nu\alpha\beta}$ is defined through $\epsilon^{0123} = 1$. The product of two Lorentz vectors is denoted by $ab = a_\mu b^\mu = a^0 b^0 - \boldsymbol{ab}$.

The Dirac matrices are denoted by γ^μ and

$$\gamma_5 = -i\gamma_0\gamma_1\gamma_2\gamma_3 = -\frac{i}{4!}\,\epsilon_{\mu\nu\alpha\beta}\gamma^\mu\gamma^\nu\gamma^\alpha\gamma^\beta\ .$$

We use the short-hand notation $\hat{a} = a_\mu\gamma^\mu$. The Pauli matrices are denoted by σ_i, $i = 1, 2, 3$.

Unless stated explicitly, we use the system of units where the Planck constant $\hbar$ and the speed of light c are equal to 1.

A.2 Gounaris-Sakurai Parametrization of the Pion Form Factor

In this Appendix, we summarize the formulas related to Gounaris-Sakurai (GS) parametrization of the pion form factor. The form factor is written as

$$F_\pi(s) = \frac{1}{1+\beta+\gamma}\left[BW_\rho^{\mathrm{GS}}(s)\big(1+\delta\,\frac{s}{m_\omega^2}\,BW_\omega(s)\big)+\beta BW_{\rho'}^{\mathrm{GS}}(s)+\gamma BW_{\rho''}^{\mathrm{GS}}(s)\right], \tag{A.1}$$

where

$$BW_\rho^{\mathrm{GS}}(s) = \frac{m_\rho^2}{m_\rho^2 - s + f_\rho(s) - i\sqrt{s}\Gamma_\rho(s)}\left(1 + \frac{d}{m_\rho}\Gamma_\rho\right);$$

$$BW_\omega(s) = \frac{m_\omega^2}{m_\omega^2 - s + i\Gamma_\omega m_\omega}\,, \tag{A.2}$$

K. Melnikov and A. Vainshtein: *Theory of the Muon Anomalous Magnetic Moment*
STMP **216**, 173–174 (2006)

with

$$f_\rho(s) = \Gamma_\rho \frac{m_\rho^2}{p_0} \left(p(s)^2 (h(s) - h(m_\rho^2)) + (m_\rho^2 - s)p_0^2 \left. \frac{\mathrm{d}h}{\mathrm{d}s} \right|_{s=m_\rho^2} \right),$$

$$h(s) = \frac{2p(s)}{\pi\sqrt{s}} \ln \frac{\sqrt{s} + 2p(s)}{2m_\pi},$$

$$d = \frac{3}{\pi} \frac{m_\pi^2}{p_0^2} \ln \frac{m_\rho + 2p_0}{2m_\pi} + \frac{m_\rho}{2\pi p_0} - \frac{m_\pi^2 m_\rho}{\pi p_0^3},$$

(A.3)

We also use $p(s) = \sqrt{1 - 4m_\pi^2/s}$, $p_0 = p(m_\rho^2)$ and

$$\Gamma_\rho(s) = \Gamma_\rho \left(\frac{p(s)}{p_0} \right)^3 \left(\frac{m_\rho^2}{s} \right)^{1/2}.$$

(A.4)

All the parameters involved in the Gounaris-Sakurai parameterization of the pion form factor are extracted from the fit to the experimental data. For numerical computations, we use [1],

$$m_\rho = 773.1 \pm 0.5 \text{ MeV}, \qquad \Gamma_\rho = 148.0 \pm 0.9 \text{ MeV},$$
$$m_{\rho^-} = 775.5 \pm 0.6 \text{ MeV}, \qquad \Gamma_{\rho^-} = 148.2 \pm 0.8 \text{ MeV},$$
$$|\delta| = (2.03 \pm 0.1) \times 10^{-3}, \qquad \arg \delta = 13.0^0 \pm 2.3^0,$$
$$m_{\rho'} = 1409 \pm 12 \text{ MeV}, \qquad \Gamma_{\rho'} = 501 \pm 37 \text{ MeV},$$
$$m_{\rho''} = 1740 \pm 21 \text{ MeV}, \qquad \Gamma_{\rho''} = 235 \text{ MeV},$$

as well as $\beta = -0.167 \pm 0.006$ and $\gamma = 0.071 \pm 0.006$. For the mass and the width of ω, we adopt $m_\omega = 782.71 \pm 0.08$ MeV and $\Gamma_\omega = 8.68$ MeV.

References

1. M. Davier, Nucl. Phys. Proc. Suppl. **131**, 123 (2004); for an update, see S. Schael et al. [ALEPH Collaboration], Phys. Rept. **421**, 191 (2005).

Index

Springer Tracts in Modern Physics

Printing: Krips bv, Meppel
Binding: Stürtz, Würzburg